Environmental Justice and Sustainability in the Former Soviet Union

Urban and Industrial Environments

Series editor: Robert Gottlieb, Henry R. Luce Professor of Urban and Environmental Policy, Occidental College

For a complete list of books published in this series, please see the back of the book.

Environmental Justice and Sustainability in the Former Soviet Union

edited by Julian Agyeman and
Yelena Ogneva-Himmelberger

The MIT Press
Cambridge, Massachusetts
London, England

This book was set in Sabon on 3B2 by Asco Typesetters, Hong Kong.
Printed on recycled paper and bound in the United States of America.

Library of Congress Cataloging-in-Publication Data

Environmental justice and sustainability in the former Soviet Union / edited by Julian Agyeman and Yelena Ogneva-Himmelberger.
    p.   cm. — (Urban and industrial environments)
Includes bibliographical references and index.
ISBN 978-0-262-01266-9 (hardcover : alk. paper) — ISBN 978-0-262-51233-6 (pbk. : alk. paper) 1. Environmental degradation—Former Soviet republics. 2. Environmental justice—Former Soviet republics. 3. Environmental policy—Former Soviet republics. 4. Former Soviet republics—Environmental conditions. I. Agyeman, Julian. II. Ogneva-Himmelberger, Yelena.
GE160.F6E576   2009
363.700947—dc22   2008042144

10   9   8   7   6   5   4   3   2   1

Contents

Contributors

Julian Agyeman   Tufts University

Caroline Campbell   Tufts University

Susan A. Crate   George Mason University

Brian Donahoe   Max Planck Institute for Social Anthropology, Germany

Jessica K. Graybill   Colgate University

Mati Heidmets   Tallinn University, Estonia

Laura A. Henry   Bowdoin College

Jüri Kruusvall   Tallinn University, Estonia

Katherine Metzo   University of North Carolina at Charlotte

Yelena Ogneva-Himmelberger   Clark University

Shannon O'Lear   University of Kansas

Tamara Steger   Central European University, Hungary

Dominic Stucker   University for Peace, Costa Rica

Maaris Raudsepp   Tallinn University, Estonia

Kate Watters   Crude Accountability

Introduction

Julian Agyeman, Yelena Ogneva-Himmelberger, and Caroline Campbell,
with additional research by Julia Prange

Although President Mikhail Gorbachev had initiated political, social, and environmental reforms in the 1980s, the corruption and economic chaos that followed the collapse of the former Soviet Union (FSU) in 1991 effectively spiraled its fifteen constituent republics into recession (Edelstein 2007, 3). The transition from Communist rule to independence and market economies resulted in a period both of intense economic and political turmoil within the countries and of tensions between them. Severe economic failure through the mid-1990s, followed by a slow recovery, the strains of decentralizing and of establishing national political power, eroding social and health care systems, and increasingly evident crises in ecological, environmental, and public health devastated these societies, at least in the short term. Although, once established, their market economies should eventually result in increased income and improved public health—greater disease prevention, higher-quality health care, healthier lifestyles, and improved regulation of environmental and occupational risks (Adeyi et al. 1997)—the deleterious impacts of the Soviet Union's collapse persist.

Even before the Berlin Wall came down in 1989, Russian society resounded with reform and experimentation. President Gorbachev successfully introduced the nation to a new period of openness, known as "glasnost," under which the closed societies of the Cold War era could be dismantled and a new global dynamic could emerge. Abandoning the restraints of state-controlled production, the Soviet regime opened doors to political and social reform it could never have touched during the Cold War. At the same time, glasnost exposed Russia's legacy of ecocidal contamination, only glimpsed during the 1986 Chernobyl disaster. Unfortunately, the fragmenting of the Soviet Union pushed Russia backward into a regressive phase of its transition, stalling reform efforts and obscuring the path toward sustainability (Edelstein 2007, 3).

Perhaps the largest challenge in assessing the political, sociocultural, and environmental aspects of the former Soviet Union is its vast geographic reach and the broad diversity of its cultural, historical, ethnic, economic, political, ecological, environmental, and social characteristics. Indeed, as our chapters show, the widely differing individual and regional histories, levels of development, economic stability, environmental and ecological activism, cultural identities, and geopolitical affiliations of the FSU republics belie their common history as Soviet Socialist Republics.

The relatively highly developed Baltic countries of Lithuania, Estonia, and Latvia are today members of the European Union (EU), and ranked 43, 44, and 45 by the United Nations Development Program's Human Development Index (UNDP HDI) report of 2007.[1] By contrast, the struggling Central Asian republics of Kazakhstan, Turkmenistan, Uzbekistan, Kyrgyzstan, and Tajikistan are ranked 73, 109, 113, 116, and 122 by the same index. In addition to these Central Asian nation-states, the loose confederation of the Commonwealth of Independent States (CIS) comprises the eastern European states of Belarus, Ukraine, and Moldova; the Caucasus states of Georgia, Armenia, and Azerbaijan; and of course the entirety of the Russian Federation.

In examining characteristics of the FSU republics, particularly those in Central Asia and the Caucasus, socioeconomic problems stand out as the most obvious—unemployment, poverty, and an unstable, transitioning market economy. These issues are coupled with a decline in income and a rise in income inequality. As is well tested and proven by Western capitalist economies (the United States in particular), greater income inequality results in worse public health and welfare for the lower-income communities. Indeed, poor public health has become an increasingly dire problem for the former Soviet Union, made all the worse by the fact that a half million Russians continue to live in contaminated areas (Zykova et al. 2001). Furthermore, the legacy of Communist growth strategies at the expense of the environment, coupled with deteriorating industrial systems and a focus on economic recovery, have wreaked havoc on ecological processes, and on the environment, more generally (Saiko 2001). The desertification of the Aral Sea, the pollution of Lake Baikal, and the radioactive contamination of thousands of square miles of Byelorussia (present-day Belarus) in the aftermath of the Chernobyl nuclear accident are but three of the best-known examples of these environmental costs. Failing to deal with the costs, Russia has slowed its transition to a democratic society to the point where it is no longer clear what it is transitioning to (see Yanitsky 1996 in Edelstein 2007, 3).

The environmental catastrophes of Russia's past are consequences of what Michael Edelstein (2007) describes as a "contaminating culture." As contaminating cultures, Russia and the United States share a general disregard for the earth, its people, and biodiversity in favor of industrialization and militarization (Edelstein 2007). Involvement in the Cold War set the two nations on parallel paths of technological and social development fueled by civilian nuclear power and the widespread use of hazardous chemicals, even in food production, all at the expense of the environment. Efforts to create a sustainable future in Russia, and the former Soviet Union more generally, must confront this toxic legacy and acknowledge the ecocide of the past (Edelstein 2007, 1).

Brown and Just, Green and Sustainable, or Points In Between?

Globally, there are many emerging, some might say "converging," agendas that highlight the need to orient humanity toward more just and sustainable futures. At one end of the activist and policy spectrum is environmental justice, conceived of in the United States and part of what activists in the global South might refer to as the "brown" antipollution, antipoverty agenda, which promotes affordable housing, clean drinking water, and infrastructure planning (McGranahan and Satterthwaite 2000).[2] At the other end is sustainable development, characterized by many as a predominantly environmental or "green" agenda (Dobson 1999, 2003), focusing on reductions in greenhouse gases, waste, and traffic and the preservation of biodiversity. Bridging these opposite ends are two "middle way" agendas: the "human security" agenda, which looks toward "sustainable security," "values the environment in itself and not merely as a set of risks . . . facilitates critical integrations of state, human and environmental security, and parallels the three linked pillars of society, economy and nature central to sustainable development" (Khagram et al. 2003, 290); and the "just sustainability" agenda (Agyeman et al. 2003; Agyeman 2005), which addresses "the need to ensure a better quality of life for all, now and into the future, in a just and equitable manner, whilst living within the limits of supporting ecosystems" (Agyeman et al. 2003, 5). The four related, central concerns of just sustainability, namely,

- quality of life;
- present and future generations;
- justice and equity;
- living within ecosystem limits,

and the different positions taken on them by environmental activists and policy makers, inform the chapters of our book.

Strides have been made in the transitioning republics of the former Soviet Union to create and implement legislation to protect and improve environmental and public health. "Many observers," Brian Donahoe tells us in chapter 1, "note the progressive nature of Russia's laws regarding environmental protection and indigenous peoples' rights." However, he argues, they have been rendered ineffective by a relentless recentralization of power and a failure to implement the laws. Another key factor in creating and maintaining this situation is that, nearly twenty years after the breakup of the Soviet Union, most of the former Communist republics are still struggling to achieve economic and political stability. That being the case, activists seeking to advance brown and green agendas could well find themselves at odds with each other, competing for political, financial, and civic investment.

Central Questions

This collection of essays from a diverse range of scholars and scholarship traditions, practitioners, and activists seeks to shed light on the growing global awareness of environmental justice, sustainable development, just sustainability, and human security in the former Soviet Union. Our efforts have been motivated by two related, overarching questions:

1. To what extent are increased popular environmental awareness and associated activism driving public policy and planning in the former Soviet republics?
2. Are there emergent, separate brown (environmental justice) and green (environmentally sustainable development) agendas or are these joining together in a single just sustainability or human security agenda?

We make no claim to be comprehensive in our geographical spread, analysis, or representation of environmental justice, (environmentally) sustainable development or just sustainability and human security in the countries of the former Soviet Union. Nor do we attempt to convey a complete picture of the economic, political, sociocultural, ecological, and environmental landscape of this diverse, complex, and expansive region. Instead, we seek to begin a conversation on the growing global awareness of environmental justice, sustainable development, just sustainability, and human security, and what shape, focus, and trajectory resultant activism and public policy and planning are taking, or might take, within the countries of the former Soviet Union.

In response to these questions, four key generalizations emerge from the growing literature about the post-Soviet transitioning economies and societies:

1. The development of a strong environmental agenda both within governments and by NGOs has been largely stymied by political leaders' focus on establishing stable and functional market economies and their lack of knowledge, determination, or financial capacity to incorporate win-win economic and environmental strategies.

There are, however, more than a few exceptions to this generalization. In an article in *Demokratizatsiia*, Laura Henry (2002, 184) wrote: "Casual readers of the Western press might be surprised to discover that in spite of the steady stream of negative reports about Russian political apathy and fatalism, the Russian environmental movement is alive and active. Environmental organizations working on issues from nuclear safety to local parks can be found in each of the Russian Federation's eighty-nine constituent regions."

2. As in much of the rest of the world, there is a divide between antipoverty campaigns (the brown agenda) and environmental campaigns (the green agenda). This divide may extend to how activists within these agendas deal with issues of common concern. Thus, although several large international organizations and numerous local, national, and international NGOs embrace and encourage the concept and practice of sustainable development as the confluence of public health and welfare, environmental justice, and environmental sustainability, many other organizations separate these issues. This separation may diminish the impact and influence of a broader movement that seeks to achieve just sustainability and human security.

3. A critical focus on poverty, national security, and economic issues may be subverting a strong agenda for just sustainability. At the 60th session of the UN General Assembly in 2005, the statements made by heads of state or high-level officials were dominated by concerns about terrorism, extremism, transnational organized crime, individual and human rights abuses, struggles against corruption, peace and stability as

integral to development, and contained few references to sustainable development.[3]

4. The cultural, political, and psychological legacy of Communism has created obstacles to the democratization of civil society, which is critical to addressing issues of environmental justice and sustainable development.

The Rise of Health-Related Economic and Environmental Issues

In the first several years of transition, circumstantial, behavioral, environmental, social, and political factors collectively contributed to a sharp decline in the health and welfare of many of the former Soviet Union populations, as the growing public health literature shows (see Adeyi et al. 1997; Bobak et al. 2000; Chen et al. 1996; Craft et al. 2006; Hela-soja et al. 2006; and Little 1998). Throughout much of the former Soviet Union, contamination has given rise to the severe psychosocial impacts of "environmental turbulence" (Edelstein 2007, 186), an indicator of the disruption and changes forced upon people as they realize they are living in a contaminated environment. These changes involve both life-styles and "lifescapes" or "how people think about themselves, their health, their homes, the environment, and those whom they rely upon for help during trying situations" (Edelstein 2007, 186). Not only are contaminated places stigmatized but also the people exposed to contamination, leading to widespread prejudice toward and increased emotional stress for the victims.

Environmental injustices are evident in regions throughout Russia where nuclear facilities have been built in territories occupied by minority groups (Edelstein 2007, 199). The Mayak facility, for instance, was constructed in a region settled mostly by the Muslim Tatar and Bashkir people and descendants of people repressed and exiled under Stalin. Activists from Muslumovo recall that the Communists, taking advantage of the passive nature of the Bashkir, built Mayak with Bashkir laborers. When the facility exploded, the Russians were evacuated and resettled, but the Tatar people were left to suffer the effects of contamination. Indeed, some 4,000 of the village's original 4,500 inhabitants—mostly Tatar—remain in Muslumovo, one of the few existing villages along the banks of the contaminated Techa River. Both Tatar and Bashkir people continue to live in contaminated regions and to harvest contaminated berries and mushrooms for their livelihoods. Fearful of their environment, some are unable, others simply unwilling, to make the lifestyle changes necessary to avoid or minimize radioactive exposure.

The Soviet emphasis on job security and full employment at the expense of productivity or efficiency has made the transition to a market economy particularly challenging, especially for people with specialized technical education or with less education. Richard Pomfret (2005) found that the three most significant factors determining how much a family can afford to spend on basic goods and services are location, children, and university education. Location and education, in particular, determine what jobs are available and feasible. Individuals with higher education of a more general nature have secured higher-paying jobs, which in turn have given them greater choice in location. These factors have contributed to the unequal distribution of wealth among regions, emigration of educated and professional people from regions with fewer job opportunities, and populations in poverty characterized by high literacy and skills.

In the post-Communist era, as people lost their jobs or received pay cuts, and as income inequality increased, living standards decreased for large segments of the population, and poverty became a dire issue. Along with these circumstances, psychological and behavioral health indicators have shown that the stressors of unemployment, uncertainty about the future, and lower living standards have led to increases in unhealthy behavior such as heavy drinking and smoking. Soil and water contamination have devastated traditional farming communities, where families are faced with the dilemma of eating local food at their own risk, selling it for cash to unsuspecting customers, or losing their foundation for survival.

As reported by Elena Craft and colleagues (2006), deleterious lifestyle behaviors (drinking, smoking, poor diet, violent or reckless behavior) are responsible for a significant portion of deaths in Russia, where an estimated 30 percent of deaths are alcohol related. According to R. E. Little (1998), mortality statistics for 1993 were alarmingly higher in Russia than in the United States, with deaths from heart disease, chronic liver disease, and cirrhosis at least 70 percent higher and from malignant neoplasms 20 percent higher. Moreover, there were twice the number of motor vehicle deaths, three times the number of homicides, and nearly nine times the number of deaths from drowning, suicide, and fire. The most dramatic increase in death rates appears to have occurred in Russia and Ukraine, where overall mortality among 18- to 65-year-old men increased more than for any other cohort, in contrast to the expectation that the most vulnerable populations (children, pregnant women, and the elderly) would experience higher mortality (Adeyi et al. 1997).

Although some research indicates that socioeconomic and psychosocial factors such as those described above may outweigh environmental factors in the increase in mortality (Little 1998), there is considerable evidence that environmental and occupational hazards have significantly contributed to the worsening of overall health in the former Soviet Union. The decentralized legal and justice systems of the Russian Federation and other FSU republics have not been able to effectively enforce industrial and environmental safety regulations; what is more, they have not had the financial capacity to maintain systems or to invest in new and emerging technologies that would lessen risk. The continued use of deteriorating Communist-era industrial and nuclear facilities whose environmental and safety regimes are widely considered to be inadequate puts the public at an even greater risk (Craft et al. 2006).

Finally, another cause of the decline in health in many parts of the former Soviet Union is inequitable or nonexistent access to basic public health and welfare services due to inadequate spending and resource allocation of governments. Thanks to a combination of all these factors, the long-term prognosis is grim: increased rates of cancer, cardiovascular disease, and chronic lung disease. Furthermore, the long-term effects of increased child morbidity (stunted growth, chronic and degenerative diseases) bode ill for future generations of former Soviet Union populations, with the Caucasus and Central Asian societies being most at risk.

What Is Being Done to Address These Issues?

The 1999 revision of *The Post-Soviet Handbook* (Center for Civil Society International, 1999) lists more that 650 post-Soviet NGOs throughout the former Soviet Union and more than 145 North American NGOs managing projects in FSU societies, all working to achieve the preservation and protection of nature, people, and their natural and cultural heritage, while ensuring a safe, healthy environment and sustainable development. This invaluable resource includes mission statements, descriptions, and contact information and speaks to Henry's quote (2002) above about the significant level of civic and environmental action in Russia.

How many of these organizations are still functioning is not clear, but there are clear indications of significant civic engagement and activism in many former Soviet Union republics regarding a host of topics, including the environment, democracy and civil society, law and human rights, security and peace, social welfare, and children and youth advocacy. By browsing the Web site of Civil Society International (CSI), a Seattle, Washington–based nonprofit NGO (http://www.civilsocietyinternational .org/), one can find hundreds of examples of organizations dedicated to these issues, ranging from international NGOs to grassroots local groups, including some specifically tailored for the FSU republics. Moreover, CSI is just one of dozens of similar resources. Indeed, following links from one Web site to the next reveals the vast and greatly interwoven network of international, national, and local organizations that seem to work tirelessly on important issues that encompass both the environmental justice and sustainable development agendas. The essay by Kate Watters, executive director of Crude Accountability (http://www.crudeaccountability.org/) on community justice against "Big Oil" in Berezovka, Kazakhstan (chapter 7), is a good example of this.

Interestingly, however, there seem to be very few organizations that deliberately meld the green (environmental or sustainable development) and brown (environmental justice) platforms into a middle-way just sustainability/human security approach, as described above. Of the more than 140 organizations listed on the World Wide Web for the fifteen former Soviet Union republics, fewer than a dozen included environmental justice in their mission statements or activity descriptions.[4] In other words, the overwhelming majority of these groups focus their energy on the green, as opposed to the brown, agenda. What nine of our twelve authors do see, however, is the emergence of at least a justice-informed environmental discourse in the former Soviet Union, if not a full-fledged environmental justice or a just sustainability/human security agenda.

Established in 1988 as a voluntary association of environmental activists at city, district, and regional levels, the Socio-Ecological Union (SEU), a Russian NGO, comprises a wide range of groups, from nature clubs and radioactive pollution victim groups to research and conservation groups from Azerbaijan, Armenia, Belarus, Estonia, Georgia, Kazakhstan, Kyrgyzstan, Moldova, Russia, Tajikistan, Turkmenistan, Ukraine, Uzbekistan, and the United States. The SEU aims to maximize the "cooperation of the intellectual potential, material and financial means, and organizational possibilities of the union's members for the preservation of nature and the protection of living beings; for the protection and revival of mankind's natural and cultural heritage; for the protection of human's physical and mental health; and for guaranteeing a safe environment and sustainable development."[5] Although human and physical health, cultural heritage, and sustainable development are included in the SEU's mission statement, the focus of its member groups and organizations is predominantly on a green, environmental, or sustainable development agenda.

The environmental movement in Russia is thought to have grown out of the student nature protection organizations of the late 1950s and early 1960s (Oldfield 1999). According to Laura Henry (chapter 2), this movement is led by scientists in the natural sciences and by engineers and is considered to have played a significant role in the collapse of the Soviet Union. In a review of *A Little Corner of Freedom: Russian Nature Protection from Stalin to Gorbachev* (Weiner 1999), Elena Lioubimtseva (2000) argues that the nature protection movement was able to exert its influence as an important venue for unsupervised and unhindered political speech. Natural scientists enjoyed greater freedom than political scientists, historians, writers, and sociologists, who were closely watched by the Communist Party. Perhaps because it wasn't taken seriously, the environmental movement was able to build itself and to have a significant impact, unfettered by censorship or worse. In the last few years, however, the movement has become a target of persecution, and several of its leaders have been arrested and jailed (Yablokov 2004; Yablokov et al. 2004).

Meanwhile, the issue of human rights has become more and more intertwined with the agenda of environmental groups. Ever since 1999, information about human rights violations and environmental injustices has been collected from media sources and published in Russian as a daily digest called "Ecologiya i prava cheloveka" ("Ecology and Human Rights," http://www.seu.ru/members/ucs/eco-hr/). These reports are distributed daily around the world via e-mail subscription. Even a cursory look through the archives of the digest reveals examples of environmental injustice in almost every region of the FSU republics.

With the breakup of the Soviet Union, environmental groups began to splinter along ideological lines—conservationist, eco-political, and confrontational, for example (Oldfield 2002). Some groups inclined more toward nature appreciation (together with a romantic or spiritual influence) and less toward activism, and others were driven more by radicalism and political contention (Yanitsky 2000). All of these splinter groups focused on a green agenda, however. It wasn't until the 1990s that the broader discourse on sustainable development played a prominent role in Russian environmentalism, as Laura Henry describes in chapter 2 of this volume. Russian academics began to turn their attention to the sustainable development discourse in the early 1990s, and works of prominent Russian geographers on this subject were compiled into an edited volume to commemorate the tenth anniversary of the United Nations Conference on Environment and Development in Rio de Janeiro (Glazovsky 2002).

Environmental groups differed with respect not only to ideology but also to the nature of their members' participation and investment (local versus foreign), to the degree of their professionalism, and to the scope of their activities (specific and local—advocacy for and protection of particular parks, animal species, and populations victimized by industrial and environmental hazards—versus more generalized and regional issues; Crotty 2003; Henry 2002; Oldfield 2002). Despite their differences, however, these groups labor under many of the same impediments to robust, effective civic engagement on the local, regional, national, and international level: distrust of government, lack of awareness of the issues, lack of professional expertise, and a disconnect between foreign, external, and professional members and the local populations for whom their groups exist (Crotty 2003).

The Russian village of Muslumovo has become a hotbed for environmental activism, where volunteers of all ages inform residents of the prevailing health risks in their contaminated community. Schoolchildren are trained to measure radioactivity in milk, and farmers are informed when it is safe to hay their fields. Activists in this village, including Greenpeace Russia, seek to empower residents with knowledge necessary to survive within a radioactive environment and to engage in civic activity (Edelstein 2007, 201).

Sustainability, Public Health, and Environmental Justice

Of the many conventions, commissions, conferences, initiatives, and meetings on global and regional environmental issues among and by a range of stakeholders in the final quarter of the twentieth century, nearly all were in response to greater global concern about the environment. In the last decade or so, however, the discourse seems to have shifted to include the issues of sustainable development, public health, and environmental justice.

The United Nations Economic Commission for Europe (UNECE) Guidelines on Access to Environmental Information and Public Participation in Decision Making (adopted in 1995) and the Johannesburg Plan for the World Summit on Sustainable Development (adopted in 2002) are but two of the significant initiatives promoting sustainable development through the agency of large, multinational NGOs and the numerous commissions and programs of the United Nations—most notably, the UN Development Program (UNDP), the UN Committee for Sustainable Development (UNCSD), the UN Economic and Social Commission for Asia and Pacific (UNESCAP), the Organization for Economic Development and Cooperation (OECD), and the World Health Organization (WHO).

Perhaps more important, however, is the Aarhus Convention on Access to Information, Public Participation in Decision Making, and Access to Justice in Environmental Matters (adopted in 1998). The culmination of more than six years of work on formulating an internationally accepted policy to encourage civic engagement, access to information, and justice regarding environmental and public health matters, the convention stems from principle 10 of the Rio Declaration from the 1992 United Nations Conference on Environment and Development (UNCED):

Environmental issues are best handled with participation of all concerned citizens, at the relevant level. At the national level, each individual shall have appropriate access to information concerning the environment that is held by public authorities, including information on hazardous materials and activities in their communities, and the opportunity to participate in decision-making processes. States shall facilitate and encourage public awareness and participation by making information widely available. Effective access to judicial and administrative proceedings, including redress and remedy, shall be provided (UN 1972, chap. 1).

Although thirteen of the fifteen FSU republics—all but the Russian Federation and Uzbekistan—ratified the Aarhus Convention, it is likely that many of their disadvantaged, isolated communities are still unaware both of the convention and of their right to know. On the other hand, communities such as those on Sakhalin Island and in Tunka and Berezovka discussed in chapters 3, 5, and 7 of this volume are working with local, national, and international NGOs to realize the aims of the convention and to develop community capacity in other areas as well.

According to the 2005 UN progress report on the Millennium Development Goals for the Central Asia and Pacific region states, among the landlocked developing countries (LLDCs) of North and Central Asia, Tajikistan, Uzbekistan, Kazakhstan, and Armenia are of greatest concern. Tajikistan (with the lowest HDI in the former Soviet Union and with 61 percent of its population hungry) and Uzbekistan (with the third lowest HDI in the former Soviet Union) have increasingly high rates of malnutrition, while Armenia is among the countries having the greatest difficulty in reducing poverty.

Positive indications of progress and good intentions in former Soviet Union countries in transition include numerous bilateral and multilateral partnerships, high participation in conventions, commissions, working groups, and several UN recognitions and awards. Thus, for its 2004 report *Education for All: The Key Goal for a New Millennium*, Kazakhstan was a finalist for the UNDP Human Development Award for Excellence in Policy Analysis and Influence.[6]

Overview of the Book

In our first chapter, "The Law as a Source of Environmental Injustice in the Russian Federation," Brian Donahoe notes "the progressive nature of Russia's laws on environmental protection and indigenous peoples' rights," but emphasizes "the instability of the legal environment...; unequal access to information about the law and about changes in the law; neglect of people's procedural rights to participate in drafting laws; and implementation problems with laws" as the principal weaknesses of the Russian legal system. These weaknesses and problems, he argues, "contribute to the law being wielded more as a technology of power and control than as an instrument for environmental security, stability, and justice."

In "Thinking Globally, Limited Locally: The Russian Environmental Movement and Sustainable Development" (chapter 2), Laura Henry asks, "Why was sustainable development embraced by many Russian actors in the early 1990s? Why now, fifteen years later, are environmentalists struggling to promote the cause of sustainability?" She answers "by exploring which features of Russia's cultural, political, and economic landscape facilitate the advancement of a sustainability agenda— and which limit it." She concludes that "Russia's strategy for recovering

from the post-Soviet economic crisis and its growing political centralization make the practical application of sustainability principles less likely
in the current period and limit the effectiveness of the movement for
sustainable development. Moreover, by focusing almost exclusively on
environmental issues, at the expense of widespread economic and social
justice concerns, the movement may have limited awareness and acceptance of the concept of sustainability among the general public."

In "Places and Identities on Sakhalin Island: Situating the Emerging
Movements for 'Sustainable Sakhalin'" (chapter 3), Jessica Graybill
argues that "socioeconomic and environmental change on Sakhalin Island in the post-Soviet era is largely occurring due to multinational-led
offshore hydrocarbon development in the Sea of Okhotsk." Pointing to
"an emerging focus on environmental justice and sustainability issues,"
she shows how "different sets of actors with different interests in—and
different visions for—creating 'sustainable Sakhalin'" are raising "questions about the local socioeconomic benefits and environmental soundness of such development at all levels, from local to international." This
chorus of voices creates "a suite of roles and practices of engagement
with issues of sustainability...among different communities on the
island."

Shannon O'Lear, in "Oil Wealth, Environment and Equity in Azerbaijan" (chapter 4), asks whether there are patterns of environmental injustice in this south Caucasus state that clearly follow ethnic or economic
lines. She uses "the concept of human security as a lens through which
to examine the impact of Azerbaijan's oil wealth on conditions of daily
life for the country's populace." From the survey data she presents,
we learn that, although Azerbaijanis "are aware of and have concerns
about environmental problems associated with the oil industry, these
are eclipsed by other, daily concerns." She uses the data to "examine
correlations between economic status and environmental concern, and
between perceived environmentally related health impacts and environmental concern."

Katherine Metzo, in "Civil Society and the Debate over Pipelines in
Tunka National Park, Russia" (chapter 5), examines two pipeline proposals from the 2000–2003 period. A 2000 proposal for a natural gas
pipeline that would run to China raised few overt complaints, whereas
the introduction of the Yukos Oil proposal a year later generated local
protest against both pipelines, each of which was slated to "cut across
protected lands adjacent to Lake Baikal including 'specially protected

zones' such as the Tunka Valley, home to a national park." Metzo suggests that part of the reason "one pipeline proposal was met with local resignation and even apathy, whereas the second raised widespread concern and prompted social action . . . lay in the different perceptions of power, responsibility, and accountability that locals had about the two proposed pipelines."

Tamara Steger, in "The Role of Culture and Nationalism in Latvian Environmentalism and the Implications for Environmental Justice" (chapter 6), argues that "animated by nationalism and cultural heritage, environmental activists in Latvia set about democratizing their country and achieving its independence from the highly centralized, authoritarian regime of the former Soviet Union in 1991." Pointing to the nationalistic and cultural elements of the Latvian environmental movement during the political changes and early transition, she asserts that "the Latvian environmental movement for independence was actually a call for environmental justice," a collective demand for natural and environmental protection coinciding with the pursuit of cultural and national recognition.

In "The Fight for Community Justice against Big Oil in the Caspian Region: The Case of Berezovka, Kazakhstan" (chapter 7), Kate Watters explores the efforts of one community to counter the deleterious environmental and public health effects of industrial oil development. She notes that, after the breakup of the Soviet Union, the new Republic of Kazakhstan was quick to invite Western corporations to invest in its economy, "seeing its vast natural resources as the key to economic development, and Western investment as an alternative to the historical economic dominance of Russia." In the mid-1990s, Kazakhstan "began negotiating with transnational oil companies to develop Tengiz and Karachaganak, two of its most lucrative oil and gas fields" and Kashagan, a newly discovered field "quickly dubbed the 'largest oil find of the past twenty years.'" With development of the fields, "revenue has flowed steadily into government and corporate coffers, even as local citizens closest to the fields continue to live in dire poverty." Watters then describes the consequences for the people of Kazakhstan and beyond: "From lost agricultural jobs and environmental health problems in refinery communities to the massive hiring of local residents for temporary construction jobs in the oil industry and the increased incidence of sexually transmitted diseases in traditional communities close to transient worker camps . . . petroleum production has intensified a downward spiral of unsustainable economic development and ecological degradation not only in Kazakhstan but in the wider Caspian Sea region."

In "The Viliui Sakha of Subarctic Russia and their Struggle for Environmental Justice" (chapter 8), Susan Crate uses an anthropological case study of a northern native people to illustrate the environmental injustices prevalent in post-Soviet Russia. She contextualizes what is known about the Viliui Sakha "within relevant research on Russia's other indigenous peoples" and compares their case to that of indigenous peoples in the diamond-mining districts of Canada's Northwest Territories. "Weaving together historical data, sociocultural analyses, and ethnographic voice," she reveals "the dynamic interplay of culture, power, and the environment." Crate argues that, despite economic decline and the Russian government's selling out of the environment to promote economic growth, and despite the Viliui Sakha's refusal to frame their cause explicitly in terms of environmental justice, if civil society continues to expand and international collaboration to gather momentum, environmental justice could be on the horizon for this and other disadvantaged peoples of the Russian Federation.

Maaris Raudsepp, Mati Heidmets, and Jüri Kruusvall, in "Environmental Justice and Sustainability in Post-Soviet Estonia" (chapter 9), describe political, economic, and social changes and the transformation of natural environments in this Baltic republic over the past decade, focusing on the emergence of social and environmental justice issues and public participation in these fields. They then present the specifics and rationale of "Sustainable Estonia 21," the national strategy of sustainable development.

Finally, Dominic Stucker, in "Environmental Injustices and Unsustainable Livelihoods: Natural Capital Inaccessibility and Loss among Rural Households in Tajikistan" (chapter 10), analyzes the relationship between environmental (in)justice and natural capital. Predominantly rural and agrarian—two-thirds of its households farm or raise livestock for their livelihoods—Takjikistan is highly dependent on natural capital. Yet only 10 percent of the nation's land is arable, severely limiting access to this capital. "With nearly 70 percent of its people living below the poverty line, and only now recovering from a protracted, bloody civil war in the 1990s, Tajikistan is the poorest country in the former Soviet Union. Its unsustainable rural livelihoods and attendant poverty only serve to increase social conflict and the likelihood of renewed civil violence."

Notes

1. The United Nations Development Program's Human Development Index (UNDP HDI) is a composite measure of average achievements in three basic realms of human development: "a long and healthy life, knowledge and a decent standard of living." http://www.hdr.undp.org/en/statistics/indices/ (accessed May 12, 2008).

2. Environmental justice is based on the principle that all people have a right to be protected from environmental pollution and to live in and enjoy a clean and healthful environment. It thus entails the equal protection and meaningful involvement of all people with respect to the development, implementation, and enforcement of environmental laws, regulations, and policies and the equitable distribution of environmental benefits (Commonwealth of Massachusetts 2002, 2).

3. Based on a review of national leaders' statements from the UN General Assembly 60th Session's General Debate, http://www.un.org/webcast/ga/60/ (accessed September 15, 2008).

4. It would be unwise, however, to assume that the information available through the World Wide Web or mainstream media outlets accurately portrays the broad range of organizations actively functioning on behalf of the former Soviet Union's citizens.

5. From http://www.seu.ru/index.en.htm/ (accessed June 19, 2007).

6. From http://hdr.undp.org/en/nhdr/monitoring/awards/2007/finalists/ (accessed May 13, 2008).

References

Adeyi, O., G. Chellaraj, E. Goldstein, A. Preker, and D. Ringold. 1997. "Health Status during the Transition in Central and Eastern Europe: Development in Reverse?" *Health Policy and Planning* 12 (2): 132–145.

Agyeman, J. 2005. *Sustainable Communities and the Challenge of Environmental Justice*. New York: New York University Press.

Agyeman, J., R. D. Bullard, and B. Evans, eds. 2003. *Just Sustainabilities: Development in an Unequal World*. Cambridge, MA: MIT Press.

Bobak, M., H. Pikhart, R. Rose, C. Hertzman, and M. Marmot. 2000. "Socioeconomic Factors, Material Inequalities, and Perceived Control in Self-Rated Health: Cross-Sectional Data from Seven Post-Communist Countries." *Social Science and Medicine* 51 (9): 1343–1350.

Center for Civil Society International. 1999. *The Post-Soviet Handbook: A Guide to Grassroots Organizations and Internet Resources*. Seattle: University of Washington Press.

Chen, L. C., F. Wittgenstein, and E. McKeon. 1996. "The Upsurge of Mortality in Russia: Causes and Policy Implications." *Population and Development Review* 22 (3): 517–530.

Commonwealth of Massachusetts. 2002. *Environmental Justice Policy*. Boston. http://www.mass.gov/envir/ej/pdf/EJ_Policy_English_Full_Version.pdf/ (accessed September 15, 2008).

Craft, E. S., K. C. Donnelly, I. Neamtiu, K. M. McCarthy, E. Bruce, I. Surkova, D. Kim, I. Uhnakova, E. Gyorffy, E. Tesarova, and B. Anderson. 2006. "Prioritizing Environmental Issues around the World: Opinions from an International Central and Eastern European Environmental Health Conference." *Environmental Health Perspectives* 114 (12): 1813–1817.

Crotty, J. 2003. "Managing Civil Society: Democratisation and the Environmental Movement in a Russian Region." *Communist and Post-Communist Studies* 36 (4): 489–508.

Dobson, A. 1999. *Justice and the Environment: Conceptions of Environmental Sustainability and Dimensions of Social Justice*. Oxford: Oxford University Press.

Dobson, A. 2003. "Social Justice and Environmental Sustainability: Ne'er the Twain Shall Meet?" In J. Agyeman, R. D. Bullard, and B. Evans, eds., *Just Sustainabilities: Development in an Unequal World*, 83–95. Cambridge, MA: MIT Press.

Edelstein, Michael R. 2007. *Cultures of Contamination: Legacies of Pollution in Russia and the U.S.* Research in Social Problems and Public Policy, vol. 14. Stamford, CT: JAI Press.

Glazovsky, Nikita, ed. 2002. *Transition to Sustainable Development at the Global, Regional and Local Levels: World Experience and Problems of Russia* [*Perekhod k ustoichivomy razvitiiu: Global'nyi, regional'nyi i lokal'nyi urovni. Zarubezhnyi opyt i problemy Rossii*]. Moscow: KMK Scientific Press.

Helasoja, V., E. Lahelma, R. Prattala, A. Kasmel, J. Klumbiene, and I. Pudule. 2006. "The Sociodemographic Patterning of Health in Estonia, Latvia, Lithuania and Finland." *European Journal of Public Health* 16 (1): 8–20.

Henry, Laura. 2002. "Two Paths to a Greener Future: Environmentalism and Civil Society Development in Russia." *Demokratizatsiya* 10 (2): 184–206.

Khagram, Sanjeev, William C. Clark, and Dana Firas Raad. 2003. "From the Environment and Human Security to Sustainable Security and Development." *Journal of Human Development* 4 (2): 289–313.

Lioubimtseva, E. 2000. "A New Glance at the Historical Role of Soviet Environmentalism." *Global Ecology and Biogeography* 9 (3): 269–270.

Little, R. E. 1998. "Public Health in Central and Eastern Europe and the Role of Environmental Pollution." *Annual Review of Public Health* 19: 153–172.

McGranahan, G., and D. Satterthwaite. 2000. "Environmental Health or Ecological Sustainability? Reconciling the Brown and Green Agendas in Urban

Development." In C. Pugh, ed., *Sustainable Cities in Developing Countries*, 72–90. London: Earthscan.

Oldfield, J. 1999. "Sustainable Development for Central and Eastern Europe: Spatial Development in the European Context." *European Urban and Regional Studies* 6 (2): 188–190.

Pomfret, R. 2005. "Kazakhstan's Economy since Independence: Does the Oil Boom Offer a Second Chance for Sustainable Development?" *Europe-Asia Studies* 57 (6): 859–877.

Saiko, T. 2001. *Environmental Crises: Geographical Case Studies in Post-Socialist Eurasia*. Harlow, UK: Prentice-Hall.

United Nations (UN). 1972. Report of the United Nations Conference on the Human Environment, Stockholm, June 5–16, 1972, chap. 1. From the UN Web site: http://www.un.org/documents/ga/conf151/aconf15126-1annex1.htm/ (accessed September 15, 2008).

Weiner, D. 1999. *A Little Corner of Freedom: Russian Nature Protection from Stalin to Gorbachev*. Berkeley: University of California Press.

Yablokov, Alexei V., ed. 2004. *The Green Movement and Civil Society: Documents of 2000–2004*. Moscow: KMK Publishing House.

Yablokov, Alexei V., et al., eds. 2004. *The Green Movement and Civil Society: Ecological Rights Violations of the Citizens in Russia*. Moscow: KMK Publishing House.

Yanitsky, O. N. 1996. *The Ecological Movement in Russia [Ekologicheskoe dvizhenie v Rossii]*. Moscow: RAN.

Yanitsky, O. N. 2000. *Russian Greens in a Risk Society: A Structural Analysis*. Helsinki: Kikimora.

Zykova, I., G. Arkhanguelskaya, and I. Zvonoa. 2001. *Chernobyl and Society: Risk Assessment [Chernobyl's i sotsium: otsenka riska]*. Saint Petersburg: Russki Esfigmen.

1

The Law as a Source of Environmental Injustice in the Russian Federation

Brian Donahoe

In a speech in January 2000, just after being tapped to succeed Boris Yeltsin as president of Russia, Vladimir Putin promised to bring the abuses of the Yeltsin era under control through a "dictatorship of the law" (*diktatura zakona*). He has since invoked this phrase in many different contexts, intending it to mean that the law is the highest authority in the land, above politics, economic considerations, and the power of any single individual. But when it comes to environmental justice for Russia's indigenous peoples, the "dictatorship of the law" has come to mean the arbitrary and highly selective application of the law (cf. Dewhirst 2005), a form of *lawfare*, defined by John Comaroff as "the effort to conquer and control indigenous peoples by the coercive use of legal means" (Comaroff 2001, 306).

As several of the chapters in this volume illustrate, the main questions of environmental justice in Russia today revolve around the industrial development and transportation of valuable natural resources—mainly fossil fuels, gold, diamonds, other mineral resources, and timber. Most of these resources are located in Siberia, on lands inhabited by indigenous peoples. Industrial development negatively affects these peoples more than others, yet they do not benefit proportionately. In this sense, industrial development can be construed as a form of environmental injustice. This chapter focuses specifically on the ways Russia's legal system can either help or hinder the protection both of the lands inhabited by indigenous peoples and of their rights to those lands and the resources on them.

Many observers of Russian legislation note the progressive nature of Russia's laws regarding environmental protection and indigenous peoples' rights, but point to the overburdened and underpaid law enforcement cadre and the ensuing lack of effective enforcement as the principal weaknesses of the Russian legal system. This chapter suggests that there are several other problems with the Russian legal system that have more to do with limiting its effectiveness when it comes to protecting the environmental—and therefore the human—rights of many of Russia's less politically powerful and less well represented peoples than the more commonly discussed issue of enforcement. These issues include the instability of the legal environment, as evidenced by the rapidity with which laws change; unequal access to information about the law and about changes in the law; neglect of people's procedural rights to participate in legislative processes; and problems with the implementation of laws. Together, these factors contribute to the law being wielded more as a technology of power and control than as an instrument for environmental security, stability, and justice.

Historical Background: The Imperial Legacy

There is a large body of literature on law, legal pluralism, and legal reforms during the time of the Russian Empire, much of which agrees that there was an antagonistic relationship between the judiciary and legal institutions on the one hand, and the monarchy on the other. Richard Wortman noted that, in tsarist times, law was "an ideal and an ornament" that "was not meant to be effected" (Wortman 2005, 151; see also *Kritika* 2006, 393). Despite the sweeping reforms represented by the Court Reform of 1864, even into the twentieth century, "the forces inimical to a law-based state remained dominant in Russia" (Wortman 2005, 169).

Historically in Russia, and later the Soviet Union, the system of justice was pluralistic and characterized by a flexibility that gave preference to local institutions and informal practices. Russian imperial law "accommodated particular social institutions extant in the population. It did not homogenize them but legalized them selectively within the whole opus of imperial legislation. The law recognized and incorporated particularity..." (Burbank 2006, 402). Rights and entitlements were not "universal and equal in the liberal manner. Instead, they emerged, piecemeal and ever fluid, out of the regime's evolving efforts to integrate diverse peoples into the expanding empire and the need to farm out administrative tasks to semi-autonomous social collectives. Without creating 'rights' that were 'natural' or 'unalienable,' this did give Russia's subjects leverage for bargaining with their rulers and made possible a degree of self-rule in the local sphere" (*Kritika* 2006, 393–394). In this way, legal pluralism in imperial Russia was more of a political process of bargaining than a legal project. This is what Jane Burbank has dubbed Russia's "imperial rights regime, which was founded on the state's assignment of rights and duties to differentiated collectivities and created conditions for including even lowly subjects in basic practices of governance" (Burbank 2006, 400).

Such practices of governance and citizenship carried over into Soviet times as well, in the form of what David Anderson calls "citizenship regimes." Siberia and the Soviet Union in general are often presented as having lacked "civil society," but Anderson argues the opposite: "as good civic practice was extended to the allegorical islands of the Russian frontier, it came to be institutionalized in a socially meaningful manner. . . . [C]ivil society in Siberia was harboured within different 'citizenship regimes' which formed restricted yet significant channels for economic and political practice" (Anderson 1996, 100; cf. Alexopoulos 2006).

But this unspoken and unofficial form of civil society has been a thorn in the side of reformers since the imperial era. Attempts to dissociate the legal from the political, and to impose a uniform, homogenized legal system based on the primacy and equality of the individual existed even in imperial times, when liberal-elite reformers, whom the "diversity of the empire and its legalism could drive . . . to despair," took it upon themselves to "civilize" all subjects of the Russian Empire by trying, unsuccessfully, to introduce Russian law to non-Russians (Burbank 2006, 424). Likewise, in the heady days after the breakup of the Soviet Union, the numerous informal means of gaining access to resources were not recognized as manifestations of civil society by Western observers and consultants who were involved in designing and pushing through the reforms intended to lead to Russia's "transition" from a socialist society to a democratic society. They were seen, rather, as impediments to civil society and to a democracy based on the rule of law. In the Yeltsin era of the 1990s and the early years of Vladimir Putin's regime, there was great optimism about the potential for civil society and democracy American style, which also included a movement toward greater faith in the rule of law.[1] John and Jean Comaroff have noted that the "new" South Africa is founded on "an ideal underpinned by an almost fetishized faith in constitutionality and the rule of law" (Comaroff and Comaroff 2004, 515), and that "the rise of neoliberalism has intensified greatly the reliance on legal ways and means" (Comaroff and Comaroff 2005, 12). This shift requires that the law come to be seen as somehow above politics. It is just such an *apolitical* type of law that Putin is invoking in his call for a "dictatorship of the law."

This increased faith in the law leads people to rely on it both as a means to redress wrongs in the civil sphere and as a way of establishing and asserting rights even against the state, as a form of resistance to the state.[2] In other words, once people perceive the law to be universal and above politics, they see it as a potential "weapon of the weak" (see Eckert 2006), which can be used "to challenge both the old and new hierarchies of power" (Merry 1994, 40, in Comaroff 2001, 306).

The Legal Environment in Contemporary Russia

In the years since the collapse of the Soviet Union, the Russian Federation has shown a willingness to address the rights of indigenous peoples and to put in place a legal framework for protecting those rights in accordance with international standards.[3] For example, Article 69 of the 1993 Russian Constitution explicitly guarantees in principle the "rights of the indigenous small-numbered peoples in accordance with the universally recognized principles and norms of international law and international agreements that the Russian Federation has entered into." Yet the definition of "indigenous small-numbered peoples" and the rights that are to be guaranteed were not delineated in the constitution. Those tasks have been addressed in a series of more recent laws that are designed to clarify the legal status of the indigenous peoples of Russia and to provide them with a legal basis for asserting their rights.[4] The most important of these laws are

• "On Guarantees of the Rights of the Indigenous Small-Numbered Peoples of the Russian Federation" (Federal Law no. 82 of April 30, 1999; revised August 22, 2004—hereafter "On Guarantees...").
• "On the General Principles of Organization of Communal Enterprises (Obshchiny) of the Indigenous Small-Numbered Peoples of the North, Siberia, and the Far East of the Russian Federation" (Federal Law no. 104 of June 20, 2000; revised August 22, 2004—hereafter "On Obshchiny...").

• "On Territories of Traditional Nature Use of the Indigenous Small-Numbered Peoples of the North, Siberia, and the Far East of the Russian Federation" (Federal Law no. 49 of May 7, 2001—hereafter "On Territories...").

Although, on their face, these framework laws appear to be quite progressive with regard to indigenous minorities' rights, in fact, they suffer from a number of weaknesses that render them ineffectual.[5] First, the insistence upon the "traditional" in the working definition of "indigenous" in Russian law effectively limits indigenous peoples to a subsistence lifestyle if they are to qualify for the rights and privileges that go along with indigenous status.[6] Forcing indigenous people to pursue an idealized "traditional" way of doing things denies them the necessary flexibility to establish a viable economic base (for treatments of the concept of traditional, see Berkes 1999; Bjerkli 1996; Donahoe 2004; Donahoe and Halemba 2006; Gusfield 1967; Habeck 2005).

Second, these laws are only very general framework laws. The actual details of their implementation are supposed to be hammered out at the regional level. As a result, they are inconsistently interpreted and unevenly enforced; indeed, the lag time between the passage of a federal law and the passage of regional implementing legislation is often so long that the federal law changes before regional implementation mechanisms can be enacted (see, for example, Wiget and Balalaeva 2004). Take the case of the Tozhu (Todzhintsy-Tuvintsy), an officially recognized indigenous small-numbered people living in the northeastern quadrant of the Siberian Republic of Tyva (Tuva). As the former president of the Tozhu branch of the Russian Association of Indigenous Peoples of the North explained, in order to activate the federal framework laws within its region, the Tyva Republic had to pass implementing legislation. He had written just such legislation and submitted it to all the appropriate legislative bodies. He never heard back. In his opinion, those in power did not want to recognize the rights of the Tozhu because the natural resources in the Tozhu Raion (district) are simply too valuable for the republic to risk losing control over them.[7] It wasn't until 2004 that the republic enacted the necessary legislation for the Tozhu to claim the rights that go along with indigenous small-numbered status, even though they had been officially granted that status at the federal level in 1993. Moreover, as we shall see below, the federal framework laws were greatly weakened in 2004, just as representatives of the Tozhu were gearing up to implement them.

With regard to protecting Russia's indigenous people's rights to land, the most highly touted of the three framework laws mentioned above, and the one that has generated the most euphoria, hope, and activity, has been the law "On Territories of Traditional Nature Use." This law was intended as a mechanism whereby indigenous peoples' lands could be declared "specially protected nature territories." Doing so could delimit at least some territories and make them federally protected for the exclusive, inalienable use of the indigenous peoples (see Bicheldei 2001, 21). According to the law, such territories of traditional nature use (TTNUs) can be established at the federal, regional, or local levels, depending on which administrative level is responsible for the land in question. In practice, however, not a single territory of traditional nature use has been established since passage of the law (Murashko 2006a). TTNUs have proven virtually impossible to establish at the local or regional level because of a fundamental contradiction between the law "On Territories..." and the new Land Code, passed just five months after that law. The problem is that almost all territories that might be candidates for TTNU status are either partly or wholly on federal land; therefore, local and regional organs of power do not have the authority to transfer control over such lands to indigenous peoples (Murashko 2002, 54). Only the federal government has the authority to do so. Yet the federal body entrusted with the power to grant TTNU status, the Ministry of Economic Development and Trade, has thus far steadfastly refused to consider applications for TTNU status, citing the lack of a universal procedure and implementation mechanism in the law (Murashko 2006a).[8] In a classic catch-22, the absence of a universal implementation mechanism and procedure is an integral aspect of the logic of such framework laws, intended to force each case to be dealt with at the local and regional levels.[9] The sad irony of the law "On Territories..." is that, of the numerous TTNUs established at the regional and local levels throughout Siberia in the ten years before its passage, most were abolished precisely because they'd been established *before* passage of a law intended to legally strengthen and secure their position.

Finally, the commitment of the government to enact and enforce these laws is questionable. Their vague wording imposes no concrete commitments on the government and provides for no mechanisms of enforcement that the indigenous peoples can fall back on in the event the government fails to make good-faith efforts to protect their lands and lifestyles. For example, Article 5 of the law "On Guarantees..." lists

eleven actions that the federal government "has the right" to undertake
to protect the lifestyles of the indigenous minorities, but nowhere does it
require the federal government to undertake these actions. It has also
been suggested that the Russian legislature prefers to keep such laws
vague in order to encourage involved parties to reach informal solutions
without recourse to the law (see Stammler and Peskov 2008 for a case
study of such negotiations).[10] Indeed, Dmitry Aleksandrovich Nesanelis,
the former vice director of the Lukoil-Varandeyneftegaz oil drilling company (Lukoil's daughter company in the Nenets Autonomous Oblast), an
anthropologist by training and the person responsible for relations between this company and the indigenous Nenets people, asserted in 2003
that it was in the interests of the state to make these laws so vague as
to be unworkable. That said, Nesanelis was quick to point out that the
oil drillers themselves were interested in clear, workable laws. As a large
multinational corporation, Lukoil is concerned with its public image
with respect to the impact its activities have on indigenous peoples and
on the environment. Nesanelis said he would prefer laws that would
give them some concrete guidelines about "what exactly they have to
pay, how, and to whom."[11] Vladislav Peskov, president of the Association of Indigenous Small-Numbered Peoples of the Nenets Autonomous
Oblast, likewise said that clearer laws were needed, but felt that, for the
time being at least, it was working out well for all parties involved,
including the Nenets: "Different people need different things. Some need
land, some need money, and the informal agreements with the drillers
allow everyone to get what they really want."[12]

Such agreements, however consistent with imperial Russian and Soviet
legal precedents (as discussed above), are likely to benefit indigenous
peoples only in the short term. Having failed to assert their legal rights
when they could have, they will find in the longer term that their economically and politically more powerful partners can turn the law
against them when it behooves them to do so.

Russia in Relation to International Law

As noted above, Russia's constitution explicitly recognizes the rights of
indigenous peoples in accordance with international law. But, like the
vagueness and difficulty of implementation of the federal laws, Russia's
lack of genuine commitment to international laws relating to indigenous peoples calls into question the sincerity of its intentions. Despite

membership in the International Labor Organization, Russia has *not* ratified ILO 169, which explicitly and unequivocally asserts the right to self-determination for all indigenous peoples. Russia's principal objection is to the term *self-determination (samoopredelenie)*, which all Russian laws on the rights of the small-numbered peoples studiously avoid, preferring instead the watered-down term *self-administration (samoupravlenie)*. This allows the Russian Federation to continue to deny indigenous peoples true control over their economic resources.

Russia is also a member of the United Nations, whose charter somewhat vaguely states that one of the purposes of the organization is "to develop friendly relations among nations based on respect for the principle of equal rights and self-determination of peoples" (Article 1, paragraph 2). Article 1 of the UN International Covenants on Civil and Political Rights and on Economic, Social and Cultural Rights (both of which Russia has ratified) more specifically asserts that "All peoples have the right to self-determination. By virtue of that right they freely determine their political status and freely pursue their economic, social and cultural development." But just as the definition of "indigenous" is contested, so is there much heated debate over who qualifies as "peoples." The United Nations has restricted the term *peoples* to those who already have "full membership in the community of nation-states" and excludes from it those who are "stateless" (Niezen 2003, 13). The United Nations recognizes indigenous peoples of *classically* colonized lands—namely, colonized lands that lie *across an ocean* from the colonizing country (the "salt-water test"; see Magnarella 2001, 2002; Niezen 2003, 138)—but has carefully avoided recognizing indigenous minorities who are not separated from their colonizers by an ocean as "peoples." This lack of recognition implicitly denies such indigenous peoples the right to self-determination—one of the arguments Russia uses to justify *not* complying with UN treaties in the case of the indigenous peoples of Siberia.

Russia is nothing if not consistent and forthright in its opposition to recent international efforts to fortify the rights of indigenous peoples. It opposed and has still not acceded to the 1998 Aarhus Convention on Access to Information, Public Participation in Decision Making, and Access to Justice in Environmental Matters. And it was, along with the United States, Canada, Australia, and New Zealand, one of the most vocal opponents of the UN Declaration on the Rights of Indigenous Peoples, which was finally passed by the UN General Assembly in September 2007, although not ratified by Russia.

Instability of the Legal Environment

The three framework laws discussed above appeared to be a decisive step toward establishing a freestanding framework to address the protection of indigenous peoples' rights as a unique and specific object of legal consideration. Recent legislation has undermined that framework, however, steadily removing the rights of indigenous peoples from special consideration. For example, Article 8, paragraph 1.1 of the law "On Guarantees..." stipulates that indigenous small-numbered peoples have the right "free of charge to possess and use, in places of their traditional habitation and economic activities, lands of various categories necessary for the realization of their traditional occupations." Article 11 of the law "On Territories..." reiterates that right by allowing for "free-of-charge use [*bezvozmezdnoe pol'zovanie*]" of land within the borders of an officially recognized "territory of traditional nature use" to representatives of the indigenous group that established the territory. But no sooner had the president's signature dried on the law "On Territories..." than the new Land Code was enacted on October 25, 2001. According to legal expert Ol'ga Murashko, the new codex effectively nullifies Article 11 of the law "On Territories..." by not explicitly recognizing the right to "free-of-charge use" of land by individuals belonging to recognized indigenous, small-numbered peoples (Murashko 2002, 54; Murashko 2004; see also Tishkov n.d.). Russia's legal environment "requires the explicit statement of rights, otherwise the prerogative is presumed to belong to the state" (Wiget and Balalaeva 2004, 134). This is in stark contrast to the United States, for example, where "powers not delegated to the United States by the Constitution, nor prohibited by it to the States, are reserved to the States respectively, or to the people" (Tenth Amendment).

Further weakening the effectiveness of the laws protecting the rights of indigenous peoples in Russia are various omnibus bills that get pushed through parliament, often at the end of a legislative session and without much debate, discussion, or publicity. Perhaps the best example is Federal Law no. 122 of August 22, 2004, with the ominous and unwieldy title: "On the Insertion of Changes in Legislative Acts of the Russian Federation and the Declaration of the Nullification of Some Legislative Acts of the Russian Federation in Connection with the Passing of the Federal Laws 'On the Insertion of Changes and Addenda in Federal Law'; 'On General Principles of the Organization of Legislative

(Representative) and Executive Organs of State Power of the Subjects of the Russian Federation'; and 'On General Principles of the Organization of Local Self-Administration in the Russian Federation.'" Article 156, paragraph 38 of this law completely abolishes the law "On the Fundamentals of State Regulation of Socioeconomic Development of the North of the Russian Federation," which established the rights of all peoples, indigenous and nonindigenous, living in regions of the "Far North or regions equivalent to the Far North" to many of the subsidies and privileges to which they have become accustomed.[13] In addition, it eviscerates two of the three framework laws mentioned above. For example, Federal Law no. 122 removes many of the rights and responsibilities of administrations at the federal, regional, and local levels to protect the rights and lands of indigenous peoples that were established in the law "On Guarantees..." (see Federal Law no. 122, article 119). It also removes from the *obshchiny* tax exemptions and powers of local self-administration that had been enshrined by the law "On Obshchiny..." (see Federal Law no. 122, article 130).

Another large omnibus bill, "On the Insertion of Changes into the Town-Planning Code and Other Legal Acts of the Russian Federation" (Federal Law no. 232 of December 18, 2006), was written by members of the pro-Putin United Russia (Yedinaia Rossiia) Party and pushed through parliament at the end of 2006. The stated purpose of the bill, which in effect represents a new town-planning codex, was to encourage housing construction by removing certain administrative obstacles to the construction of new housing (Kreindlin 2006; Ponomareva 2006). In fact, the political impetus behind Federal Law no. 232 was to pave the way for the extensive construction required for hosting the Winter Olympics in Sochi in 2014 (see *Mirovaia Energetika* 2007). In the event, however, lawmakers took this opportunity to weaken a number of laws that touch on far more than the construction of new housing, including the federal laws "On the Environmental Impact Assessment" (Federal Law no. 174 of November 23, 1995), "On the Protection of the Environment" (Federal Law no. 7 of January 19, 2002), and "On Objects of Cultural Heritage (Historical and Cultural Monuments) of the Peoples of the Russian Federation" (Federal Law no. 73 of June 25, 2002).

Most devastating to environmental justice for Russia's indigenous peoples are the changes that Federal Law no. 232 makes in the law "On the Environmental Impact Assessment" (for a thorough commentary on this earlier law, see Zlotnikova 2006). The "state environmental review [*gosudarstvennaia ekologicheskaia ekspertiza*]" is a thorough, multistage process that starts with the initial proposal of a project. An integral and indispensable early stage of this process is the "environmental impact assessment [*otsenka vozdeistviia na okruzhaiushchuiu sredu*]" (hereafter OVOS), which is a scientific report commissioned by the applicant for the project, and which must be included in the project documentation that is submitted for independent evaluation by government experts. By law, the OVOS must also be available to the public, and a public hearing must be scheduled to discuss it. In this way, the OVOS has allowed public participation and input into the development of large-scale projects likely to affect the social and natural environment of people living in the area of the projects. Although far from perfect, the law "On the Environmental Impact Assessment" has nevertheless proven to be an effective mechanism for protecting the lands where indigenous peoples live from the negative effects of industrial development.

The original law includes long and comprehensive lists of activities for which an environmental impact assessment is obligatory (*obiazatel'naia*) at the federal level (Article 11), and at the regional level (Article 12). The new revision of the environmental impact assessment law removes the word "obligatory," while the new Land-Planning Code goes so far as to allow only a "state impact assessment of the project documentation," as specified in the Land-Planning Code, and no other type. This seemingly innocuous change in effect eliminates the requirement for the OVOS or any other kind of environmental assessment at the earlier stages of the project development. The law now requires an assessment only of the *documentation* the developer provides at a later stage, but does not require an *independent* impact assessment of the actual project nor public participation in the evaluation process.

Moreover, the original law "On the Environmental Impact Assessment" notes that the purpose of the state environmental impact assessment is to forestall the negative impacts of projects on the natural environment and "the social, economic, and other consequences connected with these impacts." The Russian Association of Indigenous Peoples of the North (RAIPON) had been relying on this definition to justify calls for a new law on *ethnological* impact review (*etnologicheskaia ekspertiza*), basically as a way of extending the protections of the original law to include the impact of development projects on indigenous peoples (see Murashko 2006b). Federal Law no. 232, however, changes the very definition of environmental impact review, removing the clause on social, economic, and other consequences, thereby limiting environmental impact review to the natural environment only and nullifying an important part of the legal justification for the proposed law on ethnological impact review.

The changes introduced by both Federal Law no. 122 and Federal Law no. 232, as well as a number of other omnibus bills not discussed here, have effectively eviscerated several important legal mechanisms that indigenous peoples had relied on to protect their lands and to achieve a degree of environmental justice in the Russian Federation.

Privatization

The federal government's monopoly over the law can be best illustrated by the negotiations over the new Land Code (Zemel'nyi Kodeks; Federal Law no. 136 of October 25, 2001) and Forest Code (Lesnoi Kodeks; Federal Law no. 200 of December 4, 2006). The impact of the Land Code on the law "On Territories..." has already been mentioned above. Approximately 70 percent of all land throughout Russia is categorized as "forest fund [*lesnoi fond*]" land and has up to now been ineligible for sale. However, in an effort to stimulate development of the timber industry, the Forest Code has been rewritten to allow outright sale of formerly protected forest lands to the highest bidder, or the leasing of forest lands for two consecutive forty-nine-year periods (a form of de facto privatization). After more than three years of wrangling, the law was finally approved by parliament and signed by Putin in the first week of December 2006.

These codes are the two major legislative instruments instituting the privatization of state land. In practice, small-scale buying and selling of land has been going on since the early 1990s, mostly in and around urban areas. But the passage of the Land Code and now the Forest Code renders enormous tracts of land throughout Russia eligible for purchase by private firms and individuals, thus threatens the insecure tenure of Russia's indigenous peoples. A comparison of the sheer size and degree of specificity of these codes in relation to the three framework laws for indigenous peoples' rights is instructive. As originally written, before passage of Federal Law no. 122, the law "On Guarantees..." had 16 articles and 2,900 words, while the law "On Obshchiny..." had 24 articles and 3,300 words. The law "On Territories ..." has 18 articles and only 1,400 words. In contrast, the Land Code runs to 103 articles and 21,000 words, and the Forest Code to 109 articles and 13,000 words. The point of this counting exercise is to throw into sharp contrast the explicitness and plethora of detail with which the two codes treat the rights of landowners and lessees, on the one hand, and the vagueness and paucity of detail with which the framework laws treat indigenous peoples' rights, on the other. As noted above, Russian law requires that rights be explicitly defined and stated in order to be legally binding. Because both codes lack any explicit provision whatsoever for guaranteeing the rights of indigenous peoples to ownership or effective control over land and resources, we can assume that such rights simply do not exist as far as these laws are concerned.[14] Finally, consistent with other centralizing administrative measures that became the hallmark of the Putin regime, the Forest Code concentrates power over the management of forest lands in the hands of the federal government, and effectively removes the power of regional governments (republics, *oblasti, kraia, okrugi,* etc.) to exert control over these lands.

This puts indigenous reindeer herders and hunters at particular risk. They operate in a virtually noncash economy and could not possibly afford to purchase or lease the extensive tracts of land necessary to migrate seasonally, which is crucial both to reindeer husbandry and to the effective exploitation of wild animal resources. Landowners and lessees could also prevent them from hunting and grazing their herds on the newly privatized lands. Where post-Soviet land reform policies of privatization and exclusive land tenure have been instituted, they have favored people with financial resources and political connections, and have tended to disenfranchise politically and economically disadvantaged groups, especially those leading subsistence lifestyles.

These changes will render the southern edge of Siberia's vast boreal forest especially vulnerable to excessive timber extraction. This heavily forested region is bisected by the Trans-Siberian and Baikal-Amur railroads and by the population centers that have grown up along these railroads. It is also closest to China and Mongolia, both of which are accessible by rail and have a great demand for raw timber. Added to this, plant biomass productivity is low and tree growth extremely slow in southern Siberia (as throughout Russia), factors that make official forecasts of the forest's capacity for regeneration sound unreasonably optimistic.

The threat these changes to the Forest Code pose to the indigenous peoples of southern Siberia is best exemplified by the situation of the

Tozhu in northeastern Tyva. Nearly 90 percent of the Tozhu Raion (4,040,040 of the *raion*'s total 4,475,750 hectares; 9,984,150 of 11,060,920 acres) is officially categorized as "state forest fund lands." Until 2001, there were four nominally indigenous-led reindeer-herding *obshchiny* (cooperative enterprises) that had protected usufruct rights to approximately 1.5 million hectares (15,000 square kilometers; 3.7 million acres or 5,790 square miles) within the Tozhu Raion, but these *obshchiny* have since been dismantled. At present, the Tozhu reindeer herders and hunters range freely over much of this land, but without secure legal rights or tenure. The Tyvan government has recently announced a project to improve water transportation on the upper Yenisei River specifically to take better advantage of timber resources. The Bii Khem River, one of two large rivers that come together to form the Yenisei, runs through the southern part of the Tozhu Raion, while the Kham Syra River, a large tributary of the Bii Khem, runs through the northern part. Forests along the two rivers and their tributaries provide prime reindeer pasturage and serve as hunting and gathering grounds for the Tozhu, while the rivers themselves are important fisheries. Yet these same forests will most likely be the first to be exploited, with the timber either floated downriver or carried by barge. And because many of the forest slopes are quite steep, logging will result in severe erosion of soil into the rivers.

Compounding the threat to the Tozhu, in March 2007 the federal government approved construction of a 460-kilometer (285-mile) railroad through the Eastern Saian Mountains from Kuragino in southern Krasnoiarsk Krai through Kyzyl, the capital of Tyva, to the Elegest coalfields in central Tyva. The stated purpose of the railroad is to transport coal from the Elegest coalfields, but its proposed route goes along the western edge of the remote and inaccessible Tozhu Raion, the region of Tyva richest not only in timber but also in gold and other mineral resources. The new Forest Code, coupled with the dismantling of the *obshchiny*, will make it easier for the formerly protected forest fund lands to be sold or leased out from under the indigenous peoples of the Tozhu Raion.

Fighting Back

Russia's indigenous peoples and their representatives have demonstrated resilience and creativity in fighting back against these negative developments. One of the more often-cited cases is that of "Tkhsanom" in

Kamchatka. In early 2002, a consortium of indigenous peoples from Kamchatka applied to form a territory of traditional nature use at the federal level. The organizers received notice of refusal from the Ministry of Economic Development and Trade. However, according to the law "On Territories...," the office of the Chairman of the Russian Government (Predsedatel' Pravitel'stva R.F.) must review and decide on such applications. The consortium then applied to the Presnenskii Intermunicipal Court in Moscow, complaining of inactivity (*bezdeistvie*) on the part of the federal government. At this level, the court took a long time—nine months—and twice in that time decided (illegally, according to the Legal Center "Rodnik") to refuse to hear the complaint, and ultimately rejected it in December 2002. Rodnik tried unsuccessfully to fight this. Finally, in June 2003, Rodnik applied to the European Court of Human Rights (ECHR), complaining that the Russian government violated procedure on numerous occasions and were failing to honor the letter or spirit of their own laws by refusing to render a decision. In March 2005, the ECHR dismissed the complaint on the grounds that it did not constitute a clear violation of the UN Convention on the Protection of Human Rights and Freedom.[15]

To force the government's hand, several other applications for federal-level TTNUs (in Sakha, Taimyr, Kamchatka, and Amur) have been sent to the Chairman of the Russian Government, whose office should, by law, be responsible for deciding upon these applications.[16] These are sitting on the chairman's desk, so to speak, to make sure that the problems with implementing the law "On Territories..." are not simply swept under the rug and forgotten about.

Indigenous peoples and their representatives have demonstrated ingenuity in their attempts to assert their rights to land and resources and to protect against industrial development and extractive activities by using other laws not specifically designed for the protection of indigenous rights. One widespread strategy has been to establish national parks or other forms of specially protected nature territories (*osobo okhranaemye prirodnye territorii*) at the local or regional level or both. Federal Law no. 33 (March 14, 1995), "On Specially Protected Nature Territories [Ob osobo okhranaemykh prirodnykh territoriiakh]" explicitly empowers local and regional authorities to create such protected territories. In the Altai Republic, for example, indigenous peoples and other residents took advantage of this law to establish several locally designed and managed nature parks and, by doing so, to assert their rights to these lands.[17]

The complex and often contradictory nature of overlapping legal systems at the federal and republic levels creates confusion regarding administrative authority. The abovementioned devastating Federal Law no. 122 of August 22, 2004, removed the provision from the law "On Specially Protected Nature Territories" that allowed local and regional administrative organs to establish most categories of protected territories.[18] According to reports, forty-nine such regional-level parks in twenty-two different administrative units were threatened with liquidation by Federal Law no. 122.[19] In the event, only the Altai Republic actually moved to liquidate any parks, closing down three parks (two of them UNESCO World Heritage Sites) and threatening to close down two more, both of which had been established on the initiative of local communities.[20] But the people of the Altai did not take this sitting down and ultimately used the law to fight back. Public outcry, efforts by Greenpeace, the World Wide Fund for Nature (WWF), and other environmental organizations were all instrumental in getting Federal Law no. 122 amended. In the end, a new omnibus bill was passed that returned the law "On Specially Protected Nature Territories" to its original form, and the parks in the Altai were reinstated.[21]

Others have tried to protect land by declaring certain areas "sacred sites" (see CAFF 2004). At a seminar in the Khakass capital of Abakan in April 2002, it was suggested that a new law "On Sacred Lands of the Indigenous Peoples of the Russian Federation" should be drafted and submitted for consideration (Arbachakov 2002). Although such a law has not yet been enacted, other people have tried to use existing laws to protect sacred sites. For example, Andrew Wiget and Ol'ga Balalaeva tried invoking Federal Law no. 125 of September 26, 1997, "On Freedom of Conscience and on Religious Association," in their efforts to protect sacred places in the Khanty-Mansi Autonomous Okrug. Thus far, however, such initiatives have been unsuccessful. That law recognizes the historical priority of the Russian Orthodox Church, while "respecting Christianity, Islam, Buddhism, Judaism and other religions," but neglects to explicitly mention indigenous religions. Thus shamanism and related animistic beliefs tend to be legally categorized as "cultural" expressions rather than religious practices, so sacred shamanic sites cannot be protected under the Freedom of Conscience law.[22] Some activists have tried to turn this categorization to their advantage by invoking Federal Law no. 73 of May 24, 2002, "On Objects of Cultural Heritage (Historical and Cultural Monuments) of the Peoples of the Russian Federation," to protect sacred lands. There is, however, a catch here as well. Although the law does not explicitly exclude cultural landscapes, it is designed primarily to protect sites featuring archaeological ruins or other man-made structures of historical significance. On most sacred lands of Siberia's indigenous peoples there are no man-made structures. Moreover, labeling indigenous religious practices as "cultural" denies their religious value. Finally, as noted by Andrew Wiget, "any enrollment of sacred sites as monuments of history, culture, or nature would bring them under a state-managed protection regime, which would unilaterally exclude native people from the development and implementation of protection regimes while subjecting their religious practices to state control."[23]

A more successful effort to use the law to protect native lands also comes from the Khanty-Mansi Autonomous Okrug (KMAO). Members of the Native Assembly of the KMAO Duma asked Andrew Wiget and Ol'ga Balalaeva to design a law to guarantee their rights to continue performing the Bear Festival. Wiget and Balalaeva (2004) decided to broaden its scope a bit, and drafted an okrug-level law to protect the "folklore" of the indigenous people of Khanty-Mansi more generally. The idea was that, by protecting folklore, they would also be protecting the environment within which the folklore was embedded. It was especially important that the law should "link the perpetuation of living folklore traditions to specific communities and landscapes":

Understood in its fullest sense, it means that sacred place myths cannot exist without sacred places, nor local legends without the sites to which they are attached. In short, folklore cannot meaningfully endure if separated from the specific enculturated environment that it inhabits. Because the power to deface that environment rests with the non-native, political majority, this is potentially urgent, because KMAO is today the center of Russia's petroleum industry, and in some areas almost 90% of the land surface is licensed for petroleum production. (Wiget and Balalaeva 2004, 139–140)

Although the legislative process was slow and often frustrating, and the final version of the law is not as precise as the original and had lost some of its most important provisions, KMAO Law no. 37-03, "On the Folklore of the Native Minority Peoples of the North Living on the Territory of Khanty-Mansi Autonomous Okrug" was ultimately passed on May 30, 2003, and went into effect on June 18, 2003 (Wiget and Balalaeva 2004, 151). The most important provision with regard to protecting land from industrial development is the following: "Native Minority

Peoples living on the territory of Khanty-Mansi Autonomous Okrug are guaranteed, in the manner established by legislation:...(3) the preservation and protection of the places of the traditional circulation of folklore, and of the natural resources necessary for the perpetuation and development of folklore traditions" (KMAO Law no. 37-03, Chapter 2, article 5, paragraph 2.3).

Conclusion

Hopeful optimism about the potential efficacy of new laws led people to put their faith in the law generally. This also led to a sense of empowerment: not only were people going to appeal to the legal system to make their claims, they were going to actively participate in the creation of laws and the legislative process. Hence there was a flurry of activity throughout the 1990s and into the early years of this decade, culminating in the passage of the three framework laws discussed above. Even though the laws that ultimately passed were often watered-down versions of the originals (see, for example, Novikova 2002, 86), their passage nevertheless emboldened indigenous leaders to push for more and greater rights to their lands and resources using the formal mechanisms of the law. Regrettably, their efforts in this regard have been thwarted by the instability of particular laws, unequal access to information about changes in laws and to the legislative process, bureaucratic protocols, and legal technicalities. The state has used the law as a weapon against indigenous peoples in a form of *lawfare*, to return to the term introduced by John Comaroff, giving rise to disillusionment, alienation, and apathy and discouraging active citizenship.[24]

The weaknesses and abuses of the Yeltsin period caused many people to declare that Russia simply was not ready for democracy, and that Russia needed a strong leader in order not to dissolve into chaos. Putin declared that Russia was not ready for democracy's necessary prerequisite, the rule of law; rather, it needed what he called a "dictatorship of law." Indeed, as Richard Wortman observed in 2005: "Putin's recent policies aim at a resurgence of executive power and have shown an official contempt for the judiciary."[25]

To be fair, however, the Putin administration did much for positive judicial reform in Russia (Solomon 2005), and those efforts should be applauded. In addition, under his presidency, the Russian government moved to meet objections to at least some of its industrial initiatives.

For instance, the route of the East Siberia-Pacific oil pipeline was changed several times in response to complaints about potential ecological threats to Lake Baikal (see Metzo, chapter 5, this volume), and Putin himself objected to an earlier draft of the Forest Code because it threatened certain categories of forest.[26] However, when it came to granting indigenous peoples any sort of control over their lands and resources and any degree of real self-determination, the legal system was used first to instill the hope for real change and then to frustrate that hope.

Postscript

Although there has been a formal change of guard in Moscow with the election of Dmitrii Medvedev as president, Putin still runs the show. In 2007 a vague but much touted political platform known as "Putin's Plan" surfaced. Putin's Plan in effect guarantees continuity of Putin's policies. One of the main planks in the platform is economic growth and stability through the exploitation of natural resources. Considering that Medvedev has vowed to continue pursuing Putin's strategic vision and has installed Putin as his prime minister, there is no reason to expect that the Medvedev government will suddenly change course and implement policies that will protect indigenous peoples' rights vis-à-vis extractive industries.

Acknowledgments

This chapter started life as a joint lecture prepared with Agnieszka Halemba for the workshop Bedrohte Lebensräume: Ingidene Rechte und Erdölgewinnung in Sibirien, held in Iserlohn, Germany, August 26–28, 2005, sponsored by the Evangelische Akademie, Iserlohn. I wish to acknowledge Dr. Halemba's contribution to both the form and substance of this chapter (see Donahoe and Halemba 2006). I also thank three anonymous reviewers for constructive suggestions, and MIT Press for sensitive editing.

Notes

1. Martin Chanock recognized a similar dynamic in his study of Malawi and Zambia: "In place of a politics in which rights were delivered through the political process, now jurisprudence is the site of important decisions. No longer part of the bargaining and struggle of the political arena, decisions about entitlements

are de-politicized and rendered by means of the law" (Chanock 1985, in Comaroff and Comaroff 2005, 13).

2. Here and throughout the chapter, when I use the term *the state*, I am referring to the existing officially recognized organs of power at the federal, regional, and local levels. Although this usage tends to reify disembodied and abstracted entities, as if they existed independently of the multifarious individuals in key positions who interpret and exercise power, space does not permit a more nuanced treatment of the diverse actors and interests that make up "the state."

3. For a good overview of the types of legislation passed in the 1990s, see Arakchaa and Sumina 1999.

4. Under Russian law, place of residence (whether one lives in the "Far North" or its equivalent or on land inhabited by the indigenous small-numbered peoples) is at least as important as indigenous status for qualifying to receive state subsidies and other privileges. There is another set of laws dealing with these designations (Donahoe and Halemba 2006; Donahoe et al. 2008), but because they do not directly bear on rights to land, they will not be addressed here.

5. These laws have been thoroughly analyzed and discussed in a variety of publications in English and Russian (see, for example, Fondahl and Poelzer 2003; Novikova 1999; Kriazhkov 1996, 1999; Osherenko 2001; Mischenko 1999; Peoples Friendship University of Russia 1997).

6. In Russian law, the working definition of the term *indigenous small-numbered peoples of the North* (*korennye malochislennye narody Severa*) is "peoples living in the territories of traditional settlement of their ancestors, preserving a traditional way of life, traditional economic system and economic activities, numbering fewer than 50,000 persons within the Russian Federation, and recognizing themselves as independent ethnic communities" Federal Law no. 82 of April 30, 1999, "On Guarantees of the Rights of the Indigenous Small-Numbered Peoples of the Russian Federation"). It was first articulated in the law "On the Fundamentals of State Regulation of Socioeconomic Development of the North of the Russian Federation" (June 19, 1996).

7. Kyzyl-ool Sangy-Badra, interview, Toora-Khem village, Tozhu Raion, Republic of Tyva, July 2002.

8. Putting the Ministry of Economic Development and Trade in charge of recognizing TTNUs is in itself a violation of the law "On Territories . . . ,." according to which it should be the office of the Chairman of the Russian Government (Predsedatel' Pravitel'stva R.F.). This is discussed in greater detail below.

9. Andrew Wiget, personal communication, 2006.

10. I thank Dr. Kirill Istomin of the Siberian Studies Center, Max Planck Institute for Social Anthropology, for this suggestion, and for permission to use the interview data that follow.

11. Remarks made to Dr. Kirill Istomin during the Third Round Table on Interaction Between Indigenous Peoples and Oil Companies in the Nenets Autonomous Area (Tretii Kruglyi Stol po Vzaimodeistviiu Korennykh Narodov i Nefte-

dobyvaiushchikh Kompanii v Nenetskom Avtonomnom Okruge), Naryan-Mar, Nenets AO, May 17, 2003.

12. Remarks made to Dr. Kirill Istomin during the Third Round Table on Interaction Between Indigenous Peoples and Oil Companies in the Nenets Autonomous Area (Tretii Kruglyi Stol po Vzaimodeistviiu Korennykh Narodov i Neftedobyvaiushchikh Kompanii v Nenetskom Avtonomnom Okruge), Naryan-Mar, Nenets AO, May 17, 2003.

13. See note 4.

14. Article 97, paragraph 5 of the Land Code reiterates the right of indigenous peoples to create territories of traditional nature use (TTNU), but for reasons already discussed earlier in this chapter, no TTNUs have been created. According to article 30 of the Forest Code, members of officially recognized indigenous groups are allowed to use timber free-of-charge for subsistence needs, while Article 48 of the Forest Code notes that "while using the forest, the protection of the endemic habitats and traditional lifestyles of the indigenous small-numbered peoples must be provided for." Neither of these rights, however, approaches ownership or any other form of substantive control over land for indigenous peoples.

15. Details of this case can be found on Rodnik's Web site: http://www.rodnikcenter.ru/.

16. At the time these applications were made, the chairman of the government was Mikhail Fradkov. As of September 2007, the chairman was Viktor Zubkov.

17. I thank Agnieszka Halemba for sharing with me her insights on the Altai example.

18. Article 47 of Federal Law no. 122 of August 22, 2004 redacts the law "On Specially Protected Nature Territories" in such a way that five of the seven categories of protected territories must be of "federal significance," meaning that they can only be established at the federal level.

19. See "V Gornom Altae kipiat strasti po voprosu sokhraneniia prirodnykh parkov [In the Altai Republic Passions Seethe over Preservation of Nature Parks]." *Regnum News Agency*, February 3, 2005. Available at: http://www.regnum.ru/news/401135.html.

20. Altai Republic Resolutions nos. 180 and 198 of December 9, 2004.

21. Article 17 of Federal Law no. 199 of December 29, 2004, revokes Article 47 of Federal Law no. 122, which is the article that amended the law "On Specially Protected Nature Territories" so that local and regional authorities did not have the power to establish parks.

22. Andrew Wiget, personal communication, 2007.

23. Andrew Wiget, personal communication, 2007.

24. The Russian government's repeated, eviscerating changes in the framework laws are reminiscent of the Zimbabwe case discussed by John Comaroff, where the government keeps changing the constitution to silence its critics and to justify just doing what it wants to do.

25. This comment comes from Wortman's response to an online discussion of his 2005 article in *Kritika*. See http://www.slavica.com/journals/kritika/ekritika _pages/ekritika_wortman1.html/.

26. "Putin Asks Duma to Make More Changes to Bill on Forests." *BBC Monitoring International Reports*, July 22, 2003.

References

Alexopoulos, Golfo. 2006. "Soviet Citizenship, More or Less: Rights, Emotions, and States of Civic Belonging." *Kritika* 7 (3): 487–528.

Anderson, David G. 1996. "Bringing Civil Society to an Uncivilized Place: Citizenship Regimes in Russia's Arctic Frontier." In C. M. Hann and E. Dunn, eds., *Civil Society: Approaches from Anthropology*, 99–120. London: Routledge.

Arakchaa, K. D., and E. N. Sumina. 1999. "Prava korennykh narodov i okhrana okruzhaiushchei sredy v Arktike [The Rights of Indigenous Peoples and Environmental Protection in the Arctic]." *Zhivaia Arktika* 1999 (2): 5–12. http://www .biodiversity.ru/publications/arctic/archive/n12/rights.html/.

Arbachakov, Aleksandr. 2002. "Novosti [News]." *Medvezhii Ugol* 4 (35). http:// ecoclub.nsu.ru/isar/mu/index.htm/.

Berkes, Fikret. 1999. "Role and Significance of 'Tradition' in Indigenous Knowledge." *Indigenous Knowledge and Development Monitor* 7 (1): 19.

Bicheldei, K. A. 2001. Comments made at the Fourth Session of the Indigenous, Small-Numbered Peoples of the North, Siberia, and the Far East of Russia, April 12–13, 2001. *Mir Korennykh Narodov—Zhivaia Arktika* [The World of the Indigenous Peoples—Living Arctic] 6–7: 19–21.

Bjerkli, Bjørn. 1996. "Land Use, Traditionalism and Rights." *Acta Borealia* 1: 3–21.

Burbank, Jane. 2006. "An Imperial Rights Regime: Law and Citizenship in the Russian Empire." *Kritika* 7 (3): 397–431.

Chanock, Martin. 1985. *Law, Custom and Social Order: The Colonial Experience in Malawi and Zambia*. Portsmouth, NH: Heinemann.

Comaroff, John L. 2001. "Colonialism, Culture, and the Law: A Forward." *Law and Social Inquiry* 26 (2): 305–314.

Comaroff, John, and Jean Comaroff. 2004. "Policing Culture, Cultural Policing: Law and Social Order in Postcolonial South Africa." *Law and Social Inquiry* 29 (3): 513–545.

Comaroff, Jean, and John Comaroff. 2005. "Ethnicity, Inc.: On indigeneity and its interpellations." http://www.cics.northwestern.edu/EthnicityINC_Comaroff .rtf.

Conservation of Arctic Flora and Fauna (CAFF). 2004. *The Conservation Value of Sacred Sites of Indigenous Peoples of the Arctic: A Case Study in Northern*

Russia. CAFF Technical Report no. 11. Akureyri, Iceland: CAFF International Secretariat.

Dewhirst, Martin. 2005. "Untitled." Online posting. Sept. 15. Johnson's Russia List. http://www.cdi.org/russia/johnson/9246-18.cfm/.

Donahoe, Brian. 2004. "A Line in the Sayans: History and Divergent Perceptions of Property among the Tozhu and Tofa of South Siberia." Ph.D. diss., Indiana University.

Donahoe, Brian, and Agnieszka Halemba. 2006. "Die indigenen Völker Sibiriens: Landrechte, Legalismus and Lebensstil." *INFOEmagazin* 19: 18–21.

Donahoe, Brian, J. O. Habeck, Agnieszka Halemba, and Istvan Santha. 2008. "Size and Place in the Construction of Indigeneity in the Russian Federation." *Current Anthropology* 49 (6).

Eckert, Julia. 2006. "From Subjects to Citizens: Legalism from Below and the Homogenisation of the Legal Sphere." *Journal of Legal Pluralism* 53–54: 45–75.

Fondahl, Gail, and Greg Poelzer. 2003. "Aboriginal Land Rights in Russia at the Beginning of the Twenty-First Century." *Polar Record* 39 (209): 111–122.

Gusfield, Joseph R. 1967. "Tradition and Modernity: Misplaced Polarities in the Study of Social Change." *American Journal of Sociology* 72 (4): 351–362.

Habeck, Joachim Otto. 2005. *What it Means to be a Herdsman: The Practice and Image of Reindeer Husbandry among the Komi of Northern Russia*. Münster: LIT.

Kreindlin, Mikhail. 2006. "Ekspert Grinpis schitaet novuiu zakonodatel'nuiu initsiativu [Greenpeace Expert Evaluates New Legislative Initiative]." *Regnum News Agency*, July 26. http://www.regnum.ru/news/679252.html/.

Kriazhkov, V. A. 1996. "Pravo korennykh malochislennykh narodov na zemli [The Rights of the Indigenous Small-Numbered Peoples to Land]." *Gosudarstvo i Pravo* 1: 61–72.

Kriazhkov, V. A., ed. (compiler). 1999. *Status Malochislennykh Narodov Rossii: Pravovye Akty* [The Status of the Small-Numbered Peoples of Russia: Legal Acts]. Moscow: Tikhomirova M.Yu.

Kritika. 2006. "Tiutchev versus Foucault? Citizenship and Subjecthood in Russian History." Editorial introduction in *Kritika* 7 (3): 391–395.

Magnarella, Paul. 2001. "Sixth Annual Tribal Sovereignty Symposium: The Evolving Right of Self-Determination of Indigenous Peoples." *Saint Thomas Law Review* 14 (2): 425–447.

Magnarella, Paul. 2002. "The Right of Self-Determination." *Anthropology News*, May, 31.

Merry, Sally. 1994. "Courts as Performances: Domestic Violence Hearings in a Hawaii Family Court." In M. Lazarus-Black and S. F. Hirsch, eds., *Contested States: Law, Hegemony and Resistance*, 35–59. New York: Routledge.

Mirovaia Energetika. 2007. "Eksperty l'iut slezy po povody eko-ekspertizy [Experts Shed Tears over Eco-Impact Assessment]." *Mirovaia Energetika* 9 (45). http://www.net.ru/?q=node/2921/.

Mischenko, V. L. 1999. "Konstitutsionnye i zakonodatel'nye osnovy zashchity prav korennykh malochislennykh narodov na traditsionnoe prirodopol'zovanie: Kontseptsiia iuristov-ekologov, zashchishchaiushchikh obshchestvennye interesy [The Constitutional and Legal Basis for Defending the Rights of Indigenous Small-Numbered Peoples to Traditional Natural Resource Use: The Concept of Lawyer-Ecologists Defending Public Interests]." In N. I. Novikova and V. A. Tishkov, eds., *Chelovek i Pravo* [Man and Law], 75–81. Moscow: ID Strategiia.

Murashko, Ol'ga. 2002. "Pochemy ne rabotaet federal'nyi zakon o territoriiakh traditsionnogo prirodopol'zovaniia? [Why is the Federal Law on Territories of Traditional Nature Use Not Working?]." *Zhivaia Arktika* 11–12: 54–57.

Murashko, Ol'ga. 2004. "Politika pravitel'stva R.F. v otnoshenii prav korennykh malochislennykh narodov v Rossii v usloviiakh reformy federal'nogo zakonodatel'stva [The Politics of the Russian Government Regarding the Rights of the Indigenous Small-Numbered Peoples in Russia under Conditions of Federal Legislative Reforms]." *Russkaia Tsivilizatsiia* 17 (8). http://www.rustrana.ru/article .php?nid=1720&sq=19,23,118,898&crypt=/.

Murashko, Ol'ga. 2006a. "Zemel'nyi vopros, ili gde teper' zhit' narodam Severa [The Land Question, or Where the Peoples of the North Are Now to Live]." *Russkaia Tsivilizatsiia* 27 (11) http://www.rustrana.ru.

Murashko, Ol'ga. 2006b. *"Etnologicheskaia ekspertiza" v Rossii i mezhdunarodnye standarty otsenki vozdeistviia proektov na korennye narody* ["Ethnological Impact Assessment" in Russia and International Standards for the Evaluation of the Impact of Projects on Indigenous Peoples]. Moscow: RAIPON.

Niezen, Ronald. 2003. *The Origins of Indigenism: Human Rights and the Politics of Identity*. Berkeley: University of California Press.

Novikova, Natalia I. 1999. "Prava cheloveka i prava korennykh narodov Severa Rossii: Garmoniia ili antagonizm? [Human Rights and the Rights of the Indigenous Peoples of the North of Russia: Harmony or Antagonism?]." In N. I. Novikova and V. A. Tishkov, eds., *Chelovek i Pravo* [Man and Law], 54–63. Moscow: ID Strategiia.

Novikova, Natalia I. 2002. "Self-Government of the Indigenous Minority Peoples of West Siberia: Analysis of Law and Practice." In Erich Kasten, ed., *People and the Land: Pathways to Reform in Post-Soviet Siberia*, 83–97. Berlin: Dietrich Reimer.

Novikova, Natalia I. 2004. "Life in Reindeer Rhythms: Customary and State Regulation." *Journal of Legal Pluralism* 49: 75–90.

Osherenko, Gail. 2001. "Indigenous Rights in Russia: Is Title to Land Essential for Cultural Survival?" *Georgetown International Environmental Law Review* 13: 695–734.

Peoples Friendship University of Russia. 1997. *Pravovoi Status Korennykh Narodov Pripoliarnykh Gosudarstv* [Legal Status of Indigenous Peoples of Circumpolar States]. Moscow.

Ponomareva, Vera. 2006. "22 noiabria Gosduma posmotrit zakonoproekt, otmeniaiushchii gosudarstvennuiu ekologicheskuiu ekspertizu [On 22 November the State Duma Will Examine a Proposed Law Changing the State Environmental Impact Assessment]." *Ekologiia i pravo*, October 24. http://www.Bellona.ru/.

Solomon, Peter. 2005. "Threats of Judicial Counterreform in Putin's Russia." Paper prepared for the international conference Commercial Law Reform in Russia and Eurasia, at the Kennan Institute, Washington, D.C., April 8–9, 2005. http://www.reec.uiuc.edu/events/Conference/ACConf/lawconf_paper/solomon.pdf/.

Stammler, Florian and Vladislav Peskov. 2008. "Building a 'Culture of Dialogue' among Stakeholders in North-West Russian Oil Extraction." *Europe-Asia Studies* 60 (5): 831–849.

Tishkov, Arkadii. n.d. "Sud'ba rossiiskogo zakonodatel'stva po zashchite prav korennykh malochislennykh narodov Severa: Kommentarii u razbitogo koryta [The Fate of Russian Legislation on the Protection of the Rights of the Indigenous Small-Numbered Peoples of the North: Commentary on the Broken Trough]." http://www.biodat.ru/doc/lib/tishkov1.htm/.

Wiget, Andrew, and Olga Balalaeva. 2004. "Culture, Commodity, and Community: Developing the Khanty-Mansi Okrug Law on Protecting Native Folklore." In Erich Kasten, ed., *Properties of Culture, Culture as Property: Pathways to Reform in Post-Soviet Siberia*, 129–158. Berlin: Dietrich Reimer.

Wirtschafter, Elise Kimerling. 2006. "Russian Legal Culture and the Rule of Law." *Kritika* 7 (1): 61–70.

Wortman, Richard. 2005. "Russian Monarchy and the Rule of Law: New Considerations of the Court Reform of 1864." *Kritika* 6 (1): 145–170.

Zlotnikova, T. V. 2006. "Federal'nyi Zakon 'Ob ekologicheskoi ekspertize' [Federal Law "On the Environmental Impact Assessment"]." *Pravo-Prirode: Rossiiskoe Ekologicheskoe Zakonodatel'stvo* [Law-For Nature: Russian Environmental Legislation] no. 151. Electronic bulletin of the Tsentr Okhrany Dikoi Prirody [Center for the Protection of Wild Nature]. http://www.biodiversity.ru/.

2

Thinking Globally, Limited Locally: The Russian Environmental Movement and Sustainable Development

Laura A. Henry

Sustainable development advocates in Russia and around the world promote an economic, environmental, and social system that "meets the needs of the present without compromising the ability of future generations to meet their own needs" (WCED 1987, 43). The concept emerged in the late 1960s, reflecting a new understanding of the interdependence of economic development, environmental protection, and human well-being. The principles of sustainable development did not exert a powerful influence on Russian politics until the early 1990s, however, when Russia's post-Soviet political and economic transition initiated a search for new styles of governance. At that time, Russia seemed to offer fertile ground for sustainability, a prospect that energized both the international environmental community and Russia's domestic environmentalists. Yet even though sustainable development has been advanced enthusiastically by Russian greens and the concept codified into a variety of Russian laws, in practice, it plays only a minor role in Russian governance, environmental and otherwise.

Why was sustainable development embraced by many Russian actors in the early 1990s? Why now, fifteen years later, are environmentalists struggling to promote the cause of sustainability? This chapter answers these questions by exploring which features of Russia's cultural, political, and economic landscape facilitate the advancement of a sustainability agenda—and which limit it. Thus Russia's strategy for recovering from the post-Soviet economic crisis and its growing political centralization make the practical application of sustainable principles less likely in the current period and limit the effectiveness of the movement for sustainable development. Moreover, by focusing almost exclusively on environmental issues, at the expense of widespread economic and social concerns, the movement may have limited the awareness and acceptance of the concept of sustainability among the general public.

Sustainable Development and Russia's Post-Soviet Crisis

The collapse of the Soviet Union in 1991—the exhaustion of its Communist ideology, the end of its command economy, the disintegration of the Union itself, and the loss of the country's superpower status—created a sudden openness to new ideologies among Russia's political elite.[1] In the mid-1990s, President Yeltsin went so far as to publicize a nationwide competition to determine a new national idea for Russia.[2] As one of the most highly mobilized movements during the perestroika period of the late 1980s and early 1990s, Russian environmentalists contributed to the collapse of the Soviet regime and to the ideological debates of the post-Soviet period. In the aftermath of the Soviet Union's disintegration, however, the movement struggled under conditions of economic and political instability that led it to seek a new mobilizing platform.

The openness of state and societal actors to new ideas coincided with the growing international prominence of the concept of sustainable development (see, for example, Meadows et al. 1972; IUCN 1980; UN 1982; WCED 1987). At the most general level, sustainable development represents an effort to unite three broad aims: economic growth, environmental protection, and social equality. The concept also encompasses a model of inclusive policy making that encourages public participation. It has emerged as a powerful framework for mobilizing environmental protection and social justice movements; it has spawned international agreements, conferences, and projects, from the 1972 United Nations Conference on the Human Environment in Stockholm to the 2002 World Summit on Sustainable Development in Johannesburg. Advocates of sustainable development argue that economic growth and environmental protection are not only compatible; they are also highly interdependent. Thus the concept holds out the possibility that developing countries can find a path to becoming both wealthy and green. For a state such as post-Soviet Russia, which was in a dire economic and environmental crisis, the principles of sustainable development suggested that the country could simultaneously develop its economy, alleviate poverty, and protect and improve its natural environment. Russia formally embraced sustainable development at the 1992 Earth Summit in Rio de Janeiro when, along with 178 other states, it signed agreements including

the Rio Declaration on the Environment and Developmen
merated the principles of sustainable development, and A
action plan for achieving sustainability.

Russia entered this global movement possessing both advantages and disadvantages for achieving the goal of sustainability. The territory of the Russian Federation is home to some of the world's last pristine wilderness areas and unique ecosystems, including the taiga forest, the steppe grasslands, and Lake Baikal, the world's largest freshwater lake. Russia contains more than one-third of the world's wildlife areas, one-fifth of the world's forest cover, and a strong network of preserved lands. Indeed, at the 2002 summit in Johannesburg, Prime Minister Mikhail Kasyanov characterized Russia as an "environmental donor" to the rest of the world.[3]

On the negative side of the ledger, however, are Russia's daunting legacy of environmental degradation from the Soviet period (Feshbach and Friendly 1992), its almost unchecked natural resource exploitation, and its ongoing demographic crisis. Russia possesses ten nuclear power plants with thirty-one reactors, half of which are considered to be at high risk of breakdown by experts, and a decaying fleet of nuclear submarines (for more detailed information, see Bellona n.d.). Before 1991, Russia produced 17 percent of the world's carbon dioxide emissions. In the 1990s, more than half of the country's surface water was judged to be polluted, and air pollutants in 204 of its urban centers exceeded maximum allowable concentrations (OECD 1999, 57, 74–75). Although Russia experienced an improvement in water and air quality and a general reduction in greenhouse gas emissions during the first post-Soviet decade, this was largely attributable to the country's industrial collapse (Kotov and Nikitina 2002).[4] Since 1999, the year that Russia's economy began to recover, however, these environmental gains have been offset by the growing extraction of natural resources, including the rapid development of the oil and gas sector. Russia has the world's largest natural gas reserves and the second largest oil reserves; in recent years, it has become the world's largest exporter of natural gas.

The state of the environment reflects just one aspect of Russia's deep post-Soviet crisis; economic and social statistics also bear witness to it. The industrial decline that somewhat improved environmental indicators eroded living standards for most Russians. During Yeltsin's first term as president (1992–96), the country's GDP fell by an estimated 40 percent, unemployment rose, and inflation wiped out the savings of many

Russians. Under the Putin administration, the economy began to grow again, averaging just over 6 percent a year since 1999, but with increasing inequality between rich and poor, and between urban and rural areas (Zubarevich 2003, 3–4; UNDPR 2005, 41–42). The social stresses of the post-Soviet crisis had significant negative effects on the health of Russia's citizens as well. In 2003, Russia's Ministry of Health issued an alarming report stating that only 34 percent of Russian children were "healthy."[5] The government announced that 70 million citizens were living on territory designated as having "negative" environmental conditions (Kuvshinova and Tertychny 2005). Life expectancy for Russian men fell from 65 years in 1989 to 59 in 2004; at the same time, the birthrate fell below replacement level. In 2003, the United Nations warned that Russia's population could decline by as much as one-third by mid-century. President Putin worried publicly about Russia's survival as a nation.

Sociologist Oleg Yanitsky (2000, 267) has characterized Russia's troubled 1990s as a period of "de-modernization." Yet, due in part to its environmental assets, developed educational and health care sectors (now under stress), and declining rates of industrial pollution when compared to the Soviet period, Russia has been awarded an average or above-average grade on several sustainability rankings. In 2008, for example, Russia was given a score of 83.9 out of 100 and ranked 28 out of 149 states on the Environmental Performance Index, which measures environmental health effects, ecosystem vitality, and natural resource management, reinforcing the idea that sustainability is an achievable goal for Russia.[6]

Thinking Globally: Sustainable Development and the Soviet Legacy

Since 1991, the principles of sustainable development (*ustoichivoe razvitie*) have attracted many individuals in the Russian government and bureaucracy as well as in the environmental movement. In part, this is a testament to the concept's multifaceted nature; it espouses goals to satisfy almost every constituency. Sustainable development can accommodate competing and even, on their face, irreconcilable objectives, such as rejection of the Soviet model, on the one hand, and return to a state-controlled economy, on the other.[7] Even the Communist Party of the Russian Federation officially adopted sustainable development as a priority in the 1990s, "interpreting it as implementation of the socialist principles, increasing planning and enforcing the state intervention in economics and politics to provide just distribution of national wealth and social equality" (Demtchouk 1998, 3).

Aside from its breadth, there are other reasons why one could expect the idea of sustainable development to flourish in Russia. The concept resonates with several preexisting belief systems. First, there is a general correspondence between the logic of Soviet planning and the principles of sustainable development in that both offer a systemic analysis of production and consumption and articulate a model of appropriate relations between economy and environment. Obviously, there are important differences as well. Sustainable development promises to address uncontrolled resource use—which many environmentalists see as the most pressing environmental issue of the post-Soviet economic system. Soviet five-year plans failed to adequately price resources or to assess externalities; they measured effective resource use by planned output quotas rather than efficiency, leading to massive waste of valuable inputs (Pryde 1991; Jancar 1987). In a sustainable approach, careful pricing of natural resources and assigning value to supposedly "free goods" such as clean air and water could create incentives to ensure an adequate supply of resources for future generations. Thus sustainable development offers Russian officials and activists a new model to fill the gaping vacancy left by the demise of economic planning (as espoused by the Soviet *apparat*) and a renewed possibility of achieving the goal of social equality, but with a dramatically improved outcome for the environment.

Even more significant than its resonance with Soviet administrative models, however, sustainable development resonates with some enduring values within the environmental community in Russia. The Russian environmental tradition encompasses both a scientific and a romantic approach to nature preservation. In the Soviet period, conservation-minded scientists were able to build a small yet vocal movement for nature protection based in universities and scientific institutions (Weiner 1999, 1988). The contemporary green movement continues to be led by many academics trained in the natural sciences at Soviet institutions of higher education (Henry 2006). These individuals value the scientific method and the role that technology can play in problem solving. Sustainable development's emphasis on data collection and analysis, risk management, and technological solutions resonates with their previous professional experience. Vladimir Zakharov, president of the Center for Russian Environmental Policy, notes another affinity, commenting, "By education I am a biologist-ecologist and for me the idea of sustainable development is very close to biology.... It is akin to 'homeostasis,' the potential of maintaining an optimal balance... at the level of the organism, the population, the ecosystem, and the biosphere."[8]

The scientific strain of the green movement coexists with a romantic one, which emphasizes human connection to the land and spirituality based on the natural world. This second strain intersects with sustainable development in the small but growing trend toward establishing self-sufficient eco-villages in the Russian countryside (Bolotova 2002).[9] The broad concept of sustainable development is able to accommodate both scientific and romantic tendencies in a style reminiscent of the philosophy of Vladimir Vernadsky, a nineteenth-century Russian scientist.[10] Vernadsky suggested that the human spirit and the environment are linked in a single "noosphere" and that spiritual progress is symbolized by the use of science, or human reason, to overcome environmental problems. His philosophy's similarity to sustainable development caught the attention of many Russians, including President Yeltsin, who issued a decree comparing sustainable development to Vernadsky's noosphere, in which "national and individual wealth will be measured by spiritual values and knowledge of man, in harmony with the environment."[11]

Ideational affinities for sustainable development in Russia were reinforced by the practical benefits connected to embracing the concept for activists and organizations associated with environmental causes. By advocating public participation in policy making, sustainable development holds out the potential to restructure state-society relations in Russia, promoting democratic practices and allowing the movement to influence environmental governance in Russia. Adopting the banner of sustainable development also allows Russian environmentalists to join a vibrant international movement after years of isolation under the Communist regime, with the associated benefits of resources, partnerships, and new opportunities for activism. For many scientists, sustainable development has offered a promising avenue for pursuing research in combination with environmental advocacy. Foreign donors provide a large share of funding for Russia's NGOs (Henderson 2003); among many of these donors, sustainable development has proved to be a priority, along with preservation of biodiversity and nuclear safety.

Significant sums have been made available for environmental projects in post-Soviet Russia. Between 1991 and 1999, the European Union's Technical Aid to the Commonwealth of Independent States (TACIS) program gave 850 million euros to former Soviet states for the environment

and nuclear safety, with the majority of the funding going to Ru
the past fifteen years, the U.S. Agency for International Develop
given millions of dollars for environmental protection and sustainable
development in Russia, much of it through subcontractors such as the
Vermont-based Institute for Sustainable Communities (ISC). The ISC dis-
bursed $10 million in its Replication of Lessons Learned program from
1996 to 2006, most recently funding a program entitled "Sustainable
Development of Model Communities on a Municipal Level in Russia,"
which focused on energy efficiency.[13] Cities such as Nizhny Novgorod,
Novgorod, and Saint Petersburg have engaged in sustainable city projects in cooperation with European partners. The United Nations Development Program (UNDP) has sponsored local Agenda 21 programs in
the Altai Republic and the Bering Sea region. Ecologia, a joint American-
Russian organization, designed a sustainable development plan based on
a nonnuclear economy for the formerly closed city of Seversk. This
project will serve as model for reintegrating Russia's other closed cities
into their regional environments.

Acting Locally: The Russian Environmental Movement

Given its ideological resonance and practical benefits, the concept of sustainable development has played an important role in the Russian environmental movement during the past fifteen years. Its principles have
been taken up by a number of loosely networked environmental NGOs,
most of which are located in regional capitals. But because of Russia's
immense size, its poor communications and transportation infrastructure, and regional divisions within its environmental governance—and in
the absence of a corresponding antipoverty or social welfare movement—
environmentalists working on sustainable development have often found
themselves fragmented and unable to act as a coherent movement.

As might be expected, the goal of sustainable development has been
most wholeheartedly adopted by the scientific wing of the movement,
particularly at the national level. The Center for Russian Environmental
Policy acts as a think tank, bringing together scientists and economists at
conferences to make policy recommendations and publishing the journal
Towards a Sustainable Russia (*Na puti k ustoichivomu razvitiiu*). In
1999, the center authored "Priorities for Russia's National Environmental Policy," which elaborated methods for integrating sustainabl(
opment into nature protection. World Wildlife Fund Russia s:

model forests to demonstrate sustainable forestry practices (see WWFR n.d.). The Moscow-based NGO Eco-Accord (Eko-Soglasie) works on projects at the intersection of trade, education, consumption, and sustainable development. Organizations like the Baltic Fund for Nature of the Saint Petersburg Society of Naturalists carry out research and make recommendations on topics related to sustainable development, such as environmental protection and agriculture and the stewardship of Russia's marine reserves.

Sustainability projects at the local level based upon the Agenda 21 agreement signed at the 1992 Earth Summit were established in some regions of Russia in the 1990s. In Saint Petersburg, the Baltic Regional Agenda 21 program began in November 1997 at the initiative of Swedish and Finnish NGOs and subsequently was led by the Russian NGO Green World. Some organizations pursued projects under the Agenda 21 umbrella in Saint Petersburg. For example, Neva Clearwater conducted "Agenda 21 on the Banks of the Neva River," a project that sponsored environmental schools, where children were taught how to monitor water quality and citizens educated about pollution. There are also a variety of small groups scattered across Russia that work on projects related to alternative energy, environmentally friendly housing, and environmental education. In general, however, sustainable development has been less influential outside of the scientific and internationally funded segments of the movement. This is unfortunate because lower levels of government tend to offer the greatest opportunities for cooperation between officials and activists.

Sustainable Development and Environmental Governance in Russia

Has the movement been successful in promoting sustainable development in environmental governance in Russia? Although isolating the effects of social activism is extremely difficult (Giugni et al. 1999), since the mid-1990s, sustainable development, with some input from NGOs, has been increasingly codified in Russian law and policies. The gap between environmental regulation on paper and in practice has limited the impact of this legislation, however (see Donahoe, chapter 1, this volume). The movement's ability to promote further implementation of sustainable practices also appears to have been limited by unfavorable features of Russia's political system and economy. This section reviews Russia's environmental legislation as it relates to sustainable development and presents environmentalists' assessment of environmental protection as currently practiced.

The 1993 Constitution of the Russian Federation guarantees: "Everyone shall have the right to a favorable environment, reliable information about its condition and compensation for the damage caused to his or her health or property by ecological violations" (Article 42). Following the Earth Summit, President Yeltsin signed a decree in 1994, "On the Russian Federation Strategy for Environmental Protection and Sustainable Development," that set the framework for incorporating sustainable development into policy making: Russia should (1) pursue environmentally sustainable development within a market economy, (2) protect the human environment, (3) restore ecosystems that suffered damage in the past, and (4) participate in solving international environmental problems. President Yeltsin's decree of April 1, 1996, "Concept for the Transition of the Russian Federation to Sustainable Development," set forth an ambitious plan that enumerated the tasks necessary for shifting to sustainable development, the regional aspects of sustainable development, and decision-making criteria, as well as indicators and stages of transition to sustainable development. In 1997, Russia passed a new Forest Code, based in part on sustainable management practices.[14] Following these actions, the Russian government received a variety of international funds to support reorientation of its natural resource industries and to increase its ability to collect data and track environmental quality and public health.[15]

This steady progress led to early optimism about the integration of sustainable development into Russian environmental governance. A 1996 government report to the United Nations on Russia's progress toward implementing Agenda 21 dramatically asserts that "the transition to sustainable development may be seen as a national idea which could unite all strata of society in the cause of Russia's rebirth." It goes on to say that "the ideas of sustainable development are very much in tune with the traditions, spirit and mentality of Russia. They can play an important role in the consolidation of Russian society and in determining the State's priorities and the direction of socioeconomic transformation" (State Environmental Protection Committee of the Russian Federation 1996, 7–8).

Despite early momentum, progress on sustainable development seemed to slow in the late 1990s. Now, more than fifteen years after sustainable development "arrived" in Russia, many environmentalists express disappointment with the concept's application. Most notably, Russia never fulfilled the recommendations of earlier legislation by formally establishing a national strategy for sustainable development. On the other hand, a number of important environmental initiatives have been adopted since the late 1990s, including the target program "Environment and Natural Resources" (set to run from 2002 to 2010); a revised federal law "On Nature Protection" (2002); and the Environmental Doctrine of the Russian Federation (2002; for more on the development of these initiatives, see Oldfield 2005, 75–80). The environmental doctrine was developed in consultation with green organizations and won particular praise from many in the environmental community for the range of topics covered and the strict environmental standards recommended. The doctrine states: "Government policies in the environmental sphere are based on the following principles: sustainable development, providing for equal attention to economic, social, and ecological conditions and recognizing the impossibility of developing human society in a degraded natural environment."[16]

Although a strong legal framework has been created to advance sustainable development in Russia, the persistent gap between these laws and their implementation has led environmentalists to argue that real progress has slowed dramatically as a result of poor planning, criminality, and incomplete policy making. Some regulations under Yeltsin designed to further progress toward sustainable development did not work as intended. The penalties charged polluters for environmental damage were outstripped by high inflation and undermined by weak enforcement, for example (Kjeldsen 2000). Corrupt practices have also blunted the effectiveness of existing policies (Gavrilov 2005, 130–131; Kotov and Nikitina 2002, 11–13). In the forest sector, Greenpeace Russia (2000) has charged that the government has been negligent in its oversight of local governments' efforts to generate revenue through unregulated, illegal export of timber, if not actually complicit in those efforts. Illegal logging is especially serious in the Far East; in Primorskii Krai, for example, 50 percent of all logging is thought to be illegal (WWFR 2007). The lack of mechanisms for translating environmental objectives into actual regulation on the ground and the failure to use existing regulations seem to be the major impediments to implementation, however. A 2005 UN report assessing Russia's progress toward the Millennium Development Goals offers the following critique: "Environment protection norms and rules are dispersed among 800 various documents, of which 80 percent have recommendatory character. A large number of violations go unpunished, available legal sanctions (high penalties, closure of environmentally harmful enterprises or facilities, legal claims by citizens and public organizations for environmental damage) tend not to be applied" (UNDPR 2005, 112).

Environmentalists have repeatedly voiced their concern about this implementation gap. As early as 1998, Aleksei Yablokov, formerly President Yeltsin's ecological security advisor, noted that, although the 1996 presidential decree offered the rough outlines of sustainable development, these "major directions" were not enough to determine actual policies or "the medium- and long-term tasks of the country in this sphere" (Yablokov 1998, 5). Maria Tysiachniouk and George McCarthy (1999, 7) find this reminiscent of the situation "under the Socialist regime when the government expressed ideals which were very humanistic in theory but were never intended to be implemented." At the second All-Russia Congress on Environmental Protection in 1999, attendees evaluated Russia's adherence to its sustainable development commitments and found it deficient. "The government's systems for environmental, public health, and radiation oversight and monitoring," the congress concluded, "are deteriorating and inadequate; the process of adopting a government strategy for sustainable development is unjustifiably delayed." It pointed to "the necessity of the soonest possible formulation and approval of a Russian Federation Government Strategy for Sustainable Development" and suggested establishing a Council on Russia's Sustainable Development under the auspices of the Federation president, a suggestion that was never acted upon.[17]

Critics charge that Russia is in fact farther away from the goal of sustainable development than it was in 1992. Indeed, by using the terms *sustainable development* and *economic development* interchangeably in his speeches, President Putin seemed not to fullly appreciate the principles behind the concept.[18] "Russia has some of the best environmental laws in the world," Aleksei Yablokov observed in 2005, but "the Putin administration's attitude is that environmental protection is a luxury for a rich country—first Russia needs to exploit its natural resources in order to grow."[19]

Local Limitations: Obstacles to Promoting Sustainable Development in Russia

Despite their numerous critiques and recommendations, Russian environmentalists have thus far not been able to pressure the government to close the sustainable development implementation gap; they have clearly faced significant limitations in pressing their demands. Even when ideational affinities and resources favor mobilization, political structures and economic incentives may still hinder the effectiveness of a sustainable development movement (Tarrow 1998).

In democratic regimes, political elites generally act in a self-interested way to satisfy social groups in order to ensure favorable publicity and reelection. Under the Putin administration, however, Russia experienced a return to semi-authoritarianism, in which regional elites were increasingly marginalized and the political opposition was undermined by laws that raised the bar for political party and NGO registration. Freedom House (2008) lowered Russia's democracy ranking from approximately 3.8 in 1997 to 5.96 in 2008, on a scale from 1 to 7, labeling the country as "not free." Because sustainable development advocates public participation and decision making at the lowest possible level (subsidiarity) as essential to its process, a centralized and undemocratic state creates an especially difficult environment for its supporters.

Other obstacles hindering the ability of Russian greens to promote sustainable development include the lack of institutional support within the government; barriers to participating in policy making; and state actors' orientation toward a natural resource model of economic development. Evidence from two recent environmental campaigns illustrates these challenges.

Institutions and Resources

Even in democracies, environmental governance rarely reaches the top of a state's policy agenda, and responsibility for the environment is usually delegated to state bodies charged with environmental protection. In Russia, however, such bodies have been repeatedly weakened or eliminated altogether since the mid-1990s and those which remain are chronically underfunded (Kotov and Nikitina 2002, 14–15). Shortly after requesting a draft strategy for sustainable development in 1996, the Yeltsin administration demoted the Ministry of Ecology and the Committee for Public Health to a newly formulated State Committee on Ecology (Goskomekologia). Then, in May 2001, President Putin dissolved the State Committee on Ecology and the State Forestry Service entirely and passed their functions on to the Ministry of Natural Resources, creating a ministry charged with exploiting and protecting the natural environment at the same time. Environmental funding, whether from the state or special budget or from industrial investment, has been substantially less than expected. In particular, sustainable development programs, such as those arising from Russia's Agenda 21 commitment, have never been fully funded.

Access to Policy Making

Russia's environmental doctrine asserts the public's right to environmental information and to participate in decisions on environmental protection and natural resource use. The 1995 Federal Law on Ecological Expertise specifically grants citizens and NGOs the right to participate in environmental inspections. Yet the United Nations Division for Sustainable Development's 2002 country profile for Russia, the most recent available, states, "The development of a dialogue and cooperation between State organs and NGOs is proceeding, but not yet on a systematic basis. The Russian Federation assumes that the lack of supplementary legislation and the lack of democratic traditions in society, as well as the defects of the system for dissemination of information, impede the more active involvement of NGOs in the decision-making process" (UNDSD 2002, 39–40). Scientific experts and NGO representatives have time and again complained about their exclusion from policy making. NGOs have criticized the government for failing to ratify the Aarhus Convention, which ensures public participation in environmental governance (Zakharov 1999, 6). At the June 2002 All-Russia Conference on Environmental Security in Moscow, attended by government representatives, scientists, and environmental activists, the final resolution charged that the "constitutional environmental rights of Russian citizens, access to ecological information, compensation for damage arising from environmental offenses, and participation in environmental protection decision making are not provided for in full."[20] Viktor Danilov-Danilian, former head of the State Committee on the Environment and current head of the Russian Academy of Sciences' Institute for Water Problems, voiced his frustration with the government's lack of interest and its reluctance to provide information about the environment for the 2002 World Summit on Sustainable Development (Chistyakova 2002, 10).

Natural Resources and Economic Growth

The final, and likely most significant, impediment to the promotion of sustainable development in Russia is the current model of economic development, in which natural resource extraction industries play a preeminent role in economic recovery. The balance among Russia's economic sectors is now strongly tilted toward the natural resource sector, where firms tend also to be heavy polluters (UNDPR 2005, 128). The World Bank (2004, 12) calculated that "natural resources constitute over 80 percent of Russia's exports, and oil and gas export revenues alone are about 20 percent of GDP." Government officials appear unconvinced that pursuing sustainable development will generate the level of economic growth they believe is necessary for Russia to achieve high living standards and to regain its international stature. Environmentalists have charged that President Putin, who referred to the export of natural resources as Russia's "natural competitive advantage,"[21] used the rhetoric of sustainable development to conceal what was in fact a more traditional path toward economic modernization, and that he permitted the state-sponsored pillaging of resources by new industrial conglomerates. In recent years, the state reconsolidated its ownership of natural resource firms, partially through acquisitions by the state-owned petroleum companies Rosneft and Gazprom, further aligning its interests with extractive industries. By pursuing this strategy, President Putin implicitly reverted to a pre–sustainable development model, in which economic growth was seen as an essential precursor for resolving social problems.

Examples from Recent Campaigns

Russian environmentalists have shown a remarkable ability to mobilize under adverse conditions, but two recent campaigns, one a failure and one a success, provide evidence of the difficulty of influencing environmental governance. In 2000, the Russian government proposed legislation to permit the import of 21,000 tons of radioactive waste for reprocessing and long-term storage in the Urals and Siberia, an activity valued at approximately $20 billion dollars over ten years. Expressing outrage, Russian environmentalists cited the negative long-term environmental and health effects of this plan. In June 2000, hundreds of environmental activists from fifty-eight regions in Russia gathered together in Moscow to oppose the legislation, but received little response from the government, even though public opinion polls showed that 93 percent of Russians also opposed the scheme (Peterson and Bielke 2001, 69). Environmentalists gathered signatures in support of a national referendum on the issue. After months of effort, however, the Central Election Commission ruled that only three-quarters of the signatures were valid, causing the campaign to fall just short of the two million signatures required. Greens charged that the signature count had been manipulated, but the legislation on importing nuclear waste passed in 2001.

More recently, environmentalists opposed the plans of Transneft, the state-owned oil transport company, to construct an oil pipeline from Siberia to the Pacific that would pass just 800 meters (less than a half mile) from Lake Baikal. Environmental activists, joined by the state's own State Ecological Expertise committee, argued that the pipeline posed a major pollution threat, but their reports were ignored by the Federal Technical Inspectorate. Then, in July 2006, after months of negative publicity in the international press, President Putin unilaterally ordered the route of the pipeline moved at least 25 miles from Baikal. Although they welcomed his decision, the greens remained deeply skeptical. "Nobody is fooled," said Vladimir Chuprov, chairman of Greenpeace Russia. "Putin's decision was good for the environment, but it was only one isolated deed which does not foreshadow any change in the government's ecological policy."[22]

Acting Globally: Seeking International Opportunities

Frustrated domestically, Russian environmentalists have increasingly sought international opportunities. For all its resistance and inaction on the domestic front, the Russian government has not abandoned the language of sustainable development and continues to participate in international conferences and agreements on the topic. For example, the Russian Federation was either a partner state or a target country for at least nine multi-stakeholder projects on sustainable development that emerged from the 2002 Johannesburg Summit.[23] During Russia's G-8 presidency in 2006, the government repeatedly addressed the issue of implementing the Kyoto Protocol and increasing the energy efficiency of the Russian economy, which is more carbon intensive than European economies. President Medvedev has reiterated Russia's intention to adddress these issues. These and similar commitments to sustainable development at the international level create opportunities for environmentalists working inside Russia to activate third-party pressure on Russia to meet its obligations. When undemocratic states disregard domestic public opinion and resist pressure for policy change, domestic actors can attempt to trigger a "boomerang effect," in which transnational actors and third states press the target state into adjusting its policies (Keck and Sikkink 1998).

Russian environmentalists have leveraged Russia's international commitments to push for progress domestically in a number of cases. For example, in the 1990s, Eco-Accord used the government's published sustainable development goals to measure its performance, holding hearings entitled "Russia and Rio+5: What Has Been Done?" and collecting comments from governmental and nongovernmental observers alike in order to compile a list of recommendations for future action. Then, just before the 2002 World Summit on Sustainable Development in Johannesburg, a group of seventeen of Russia's most prominent environmental leaders issued their "Declaration of the Russian Green Movement for the Rio+10 Summit," which stated, "Today, Russia's system of environmental protection has considerably weakened, with principles of sustainable development not taken into account in addressing objectives of economic development." The declaration proposed three goals: developing civil society, ensuring the health of the environment, and raising the value of natural resources.[24] At the same time, a number of Moscow-based environmental NGOs, including some of the most influential groups such as Greenpeace Russia and World Wildlife Fund Russia promoted the country's ratification of the Kyoto protocol, which finally occurred in 2004 (Henry and Sundstrom 2007). Aleksei Kokorin of World Wildlife Fund Russia credits his organization's international ties for allowing it to reach out to European greens, who in turn encouraged European leaders such as Tony Blair, Jacques Chirac, and Gerhard Schroeder to press President Putin on Russia's commitment to the climate change agreement.[25] Environmentalists are now using these same networks to push for Russian implementation of the Kyoto Protocol. Russian greens also have embraced international environmental governance programs, such as the Forest Stewardship Council's sustainable forest certification program, as a means of combating weak domestic laws or enforcement (Tysiachniouk 2006).

Broadening the Definition of Sustainability

Debates over the prospects for sustainable development in Russia are still limited to elite circles, despite the concept's espousal of participatory politics. Thus the scientific wing of the environmental movement is still struggling to propagate the idea of sustainable development beyond its relatively narrow networks. The concept is only beginning to capture the imagination of grassroots environmental groups and remains largely unfamiliar to the broader public, thanks in large part to the technical nature of many issues related to sustainability and the greens' inability to gain media coverage of their campaigns. Another problem, however, is the two-dimensional nature of the sustainability debate in Russia, which focuses on the environment and economy at the expense of other important concerns. Sustainability advocates have little to say about social concerns, for example, a glaring omission in light of Russia's demographic crisis and shrinking population that President Putin has declared a "national emergency" (for details on Putin's response to the crisis, see Kupchinsky 2006). A 2005 poll published in Pravda indicated that 44 percent of Russians regard the Russia's demographic problems as "critical."[26] Given the great concern within the government and the public about the crisis, including eroding health, education, and housing infrastructure, explicitly bringing social issues into the current debate on sustainable development presents a possible path toward broadening its appeal and inspiring government action.

One of Russia's green parties, the Union of Greens, has recognized that the social justice component of sustainable development holds the potential for broader mobilization. Its platform highlights the linkages between environmental degradation, health, poverty, and poor governance—a framing of sustainability more likely to capture the public's attention than one focused purely on conservation. The party's broader orientation lends itself to developing ties with other groups concerned about Russia's disintegrating social sector, including traditional Communist Party supporters such as pensioners and rural residents. The Union of Greens have struggled with the extremely difficult political conditions facing new parties, however, and the party currently operates as a faction inside the liberal democracy party Yabloko rather than as a fully independent political party.

Russia's rural regions suffer from lower living standards, but have been largely neglected by social activists. A 2003 UNDP report showed that there is great inequality in conditions across the country's regions, with Russia's Human Development Index ranging from a level equivalent to Slovenia in Moscow, for example, to the level of Gabon in the Republic of Tyva. The overall poverty rate tends to be 30–40 percent higher in rural areas and the mortality rate for children ages one to four is twice as high among rural as urban populations (Zubarevich 2003, 4; UNDPR 2005, 160). A 2005 UN report concludes, "Geographical location is the most important factor determining welfare inequality in

Russia today" (UNDPR 2005, 37). There also is some indication that these problems tend to be more severe in non–ethnic Russian regions, although the evidence is mixed (UNDPR 2005, 156; Zubarevich 2003, 25). Recent campaigns, such as environmental and social activism on Sakhalin Island (see Graybill, chapter 3, this volume), where there is a significant indigenous population, demonstrate that NGO demands for environmental and social justice can be mutually reinforcing and may attract a broader swath of the general public to the cause of sustainability. Since it has proven difficult to increase government action on sustainable development, the movement might next focus on increasing the public's demand for political change.

Conclusion

Paradoxically, the economic crisis that initially prompted the search for new models of governance in Russia has created a situation in which the institutionalization of sustainable development is less likely. It is difficult to imagine Russia integrating principles that limit natural resource exploitation when the main engine of the economy is the oil and gas sector. Political conditions also are challenging. As long as the Russian government continues on a path of centralizing political power, favoring natural resource industries, and controlling information, Russian greens will have difficulty promoting sustainable development. Without greater public access to policy making and stable, transparent, and effective governing institutions, the gap between the rhetoric and the practice of sustainable development is likely to persist.

The incorporation of sustainable development into many of Russia's environmental laws is no small achievement, however, even if these laws have not been fully implemented. Movements in many other countries, for example, have failed to push their governments beyond the use of sustainability as a political slogan. This may reflect an inherent difficulty in the concept of sustainability itself, whose broad and ambitious abstract principles require significant effort to translate into concrete progress. As long as these sustainable environmental laws exist in Russia, buttressed by opportunities for action at the international level, there is hope of gradually infusing them with practical meaning. In the 1970s, dissident intellectuals used Soviet accession to the Helsinki Accords to publicize the country's poor human rights record, indirectly leading to Gorbachev's policy of glasnost. Sustainable development may be largely a rhetorical device today, but Russia's green movement continues to work toward making it a reality in the future.

Notes

1. Scholars working on international norms have recognized that "world historical events such as wars and major depressions in the international system can lead to a search for new ideas and norms" (Finnemore and Sikkink 1998, 909).

2. For an account of Yeltsin's efforts to develop a new national ideology, see Breslauer and Dale 1997.

3. "Speech of Russian Federation Prime Minister M. M. Kasyanov," World Summit on Sustainable Development, Johannesburg, September 3, 2002. http://www.un.org/events/wssd/statements/russiaR.htm.

4. The United Nations estimates that both polluted effluents and atmospheric emissions fell by one-third during the post-Soviet period (UNDPR 2005, 109–110).

5. "Only 20 Percent of Russians Healthy; 60 Percent of Children Sick," *(Un)-Civil Societies* 4 (21).

6. The Environmental Performance Index is compiled by the Yale Center for Environmental Law and Policy and the Center for International Earth Science Information Network of Columbia University and is available at http://epi.yale.edu/Home/.

7. In interviews, Russian environmentalists have critiqued sustainable development for being too much like Communism or, alternatively, as a postindustrial value incompatible with Russian realities (Tysiachniouk and McCarthy 1999).

8. Vladimir Zakharov, director, Center for Russian Environmental Policy, interview, Moscow, June 28, 2005.

9. Many founders of eco-settlements are followers of Vladimir Megre, the author of the Ringing Cedars book series.

10. This similarity between sustainable development and Vernadsky's noosphere was mentioned several times by my Russian interviewees from the environmental movement in 1999–2001 and 2005 and is fully developed by Jonathan Oldfield and Denis J. B. Shaw (2002).

11. Presidential Decree "Concept for the Transition of the Russian Federation to Sustainable Development," April 1, 1996, 9.

12. European Commission, "Figures 91/99." http://ec.europa.eu/comm/external_relations/ceeca/tacis/.

13. For more information on the ISC-sponsored model communities program, see the Institute for Sustainable Communities Russia Web site, at http://www.iscmoscow.ru/round15/press_releasie_15.htm.

14. A more recent version of the Forest Code, passed in 2007, has been highly criticized by environmentalists for inconsistences.

15. These funds included $20.1 million for preserving biodiversity, $60 million for a reduction of the consumption of ozone-depleting substances, and $110 million for environmental management projects from the World Bank in 1996 alone. For more examples of foreign funding for sustainable development, see NCSD 1999, 8–9.

16. The Russian text of the Environmental Doctrine is available at http://www.scrf.gov.ru/documents/decree/2002_1225-r.shtml/.

17. "Resolution of the Second All-Russia Congress on Environmental Protection," *Towards a Sustainable Russia* 1 (12): 9, 10, 12.

18. Oldfield and Shaw note that when "sustainable development" is translated into Russian (*ustoichivoe razvitie*), it literally means "steady" or "stable development."

19. Aleksei Yablokov, chairman of Union of Greens/Green Russia, interview, Moscow, July 15, 2005.

20. "Resolution of the All-Russia Conference on Environmental Security," *Towards a Sustainable Russia* 20: 33.

21. Vladimir Putin, "Speech at the Ninth Saint Petersburg International Economic Forum," June 14, 2005.

22. "Russian Ecologists Despair over Lack of Impact, Government Backing," Agence France-Presse, May 8, 2006. As reprinted in *Russian Environmental Digest* 8 (20).

23. A complete project list available from the United Nations Department of Economic and Social Affairs, Division of Sustainable Development (http://www.un.org/esa/sustdev/partnerships/partnerships.htm).

24. "Declaration of the Russian Green Movement for the Rio+10 Summit," *Towards Sustainable Development* 20: 3.

25. Aleksei Kokorin, coordinator, Climate Change Program, World Wildlife Fund Russia, interview, Moscow, June 29, 2005.

26. "Demographic Crisis Poses Serious Danger to Russia's Future," *Pravda*, November 24, 2005. http://english.pravda.ru/russia/politics/24-11-2005/9286-crisis-0/.

References

Bellona Foundation. n.d. "Nuclear Russia." http://www.bellona.org/subjects/Nuclear_Russia.

Bolotova, Alla. 2002. "Ecological Settlements: Between City and Village." *Zhurnal sotsiologii i sotsial'noi antropologii* [Journal of Sociology and Social Anthropology] 1.

Breslauer, George W., and Catherine Dale. 1997. "Boris Yel'tsin and the Invention of a Russian Nation-State." *Post-Soviet Affairs* 13 (4): 303–333.

Chistyakova, E. K. 2002. "Russia and the World on the Eve of the Rio+10." *Towards a Sustainable Russia* 20: 9–10.

Demtchouk, Artour L. 1998. "Sustainable Development: New Political Philosophy for Russia?" Paper presented at the Twentieth World Congress of Philosophy, Boston, August 10–15. http://www.bu.edu/wcp/Papers/Poli/PoliDemt .htm/.

Feshbach, Murray, and Alfred Friendly, Jr. 1992. *Ecocide in the USSR: Health and Nature under Siege*. New York: Basic Books.

Finnemore, Martha, and Kathryn Sikkink. 1998. "International Norm Dynamics and Political Change." *International Organization* 52 (4): 887–917.

Freedom House. 2008. *Nations in Transit*. http://www.freedomhouse.hu/images/ fdh_galleries/NIT2008/NT-Russia-final1.pdf.

Gavrilov, V. V. 2005. "Economic Efficiency of the Model of Environmental Impact Charges." In United Nations Development Program Russia, *Russia in 2015: Development Goals and Policy Priorities*, 130–131. Moscow: United Nations Development Program.

Giugni, Marco, Doug McAdam, and Charles Tilly. 1999. *How Social Movements Matter*. Minneapolis: University of Minnesota Press.

Greenpeace Russia. 2000. "Results of the Public Opinion Poll concerning the Use and Protection of the Forests of Karelia." Moscow. February.

Henderson, Sarah. 2003. *Building Democracy in Contemporary Russia: Western Support for Grassroots Organizations*. Ithaca, NY: Cornell University Press.

Henry, Laura A. 2006. "Shaping Social Activism in Post-Soviet Russia: Leadership, Organizational Diversity, and Innovation," *Post-Soviet Affairs* 22 (2): 99–124

Henry, Laura A., and Lisa McIntosh Sundstrom. 2007. "Russia and the Kyoto Protocol: Seeking an Alignment of Interests and Image." *Global Environmental Politics* 7, 4 (November): 47–69.

International Union for Conservation of Nature (IUCN). 1980. *World Conservation Strategy*. Gland, Switzerland.

Jancar, Barbara. 1987. *Environmental Management in the Soviet Union and Yugoslavia*. Durham: Duke University Press.

Keck, Margaret E., and Kathryn Sikkink. 1998. *Activists Beyond Borders*, Ithaca: Cornell University Press.

Kjeldsen, Stig. 2000. "Financing of Environmental Protection in Russia: The Role of Charges." *Post-Soviet Geography and Economics* 41 (1): 48–62.

Kotov, Vladimir, and Elena Nikitina. 2002. "Reorganisation of Environmental Policy in Russia." Fondazione Eni Enrico Mattei Note di Lavoro Series. July. http://www.feem.it/Feem/Pub/Publications/WPapers/WP2002-057.htm.

Kupchinsky, Roman. 2006. "Russia: Tackling the Demographic Crisis." *Radio Free Europe/Radio Liberty*, May 19.

Kuvshinova, Olga, and Nikolai Tertychny. 2005. "Number of Russian Cities with High Pollution Up 50 Percent over 5 Years." ITAR-TASS, March 22. Reprinted in *Russian Environmental Digest* 7 (13).

Meadows, Donatella, Dennis L. Meadows, Jørgen Randers, and William W. Behrens III. 1972. *The Limits to Growth: A Report for the Club of Rome's Project on the Predicament of Mankind*. New York: Universe Books.

National Council for Sustainable Development (NCSD). 1999. *Sustainable Development Report: Russia*.

Oldfield, Jonathan D. 2005. *Russian Nature: Exploring the Environmental Consequences of Societal Change*. Burlington, VT: Ashgate.

Oldfield, Jonathan D., and Denis J. B. Shaw. 2002. "Revisiting Sustainable Development: Russian Cultural and Scientific Traditions and the Concept of Sustainable Development." *Area* 34 (4): 391–400.

Organization for Economic Cooperation and Development (OECD). 1999. *Environmental Performance Reviews: The Russian Federation*. Paris.

Peterson, D. J., and Eric K. Bielke. 2001. "The Reorganization of Russia's Environmental Bureaucracy: Implications and Prospects." *Post-Soviet Geography and Economics* 42 (1): 65–67.

Pryde, Philip R. 1991. *Environmental Management in the Soviet Union*. Cambridge: Cambridge University Press.

State Environmental Protection Committee of the Russian Federation. 1996. *Russian Federation Country Profile: Implementation of Agenda 21*. December. http://www.un.org/esa/earthsummit/rusia-cp.htm/.

Tarrow, Sidney. 1998. *Power in Movement: Social Movements and Contentious Politics*. 2nd ed. Cambridge: Cambridge University Press.

Tysiachniouk, Maria. 2006. "Forest Certification in Russia." In B. Cashore, F. Gale, E. Meidinger, D. Newsom, eds., *Confronting Sustainability: Forest Certification in Developing and Transitioning Countries*, 261–295. Yale School of Forestry and Environmental Studies, Report no. 8. New Haven, CT.

Tysiachniouk, Maria, and George McCarthy. 1999. "A Comparison of Attitudes of Russian and U.S. Environmentalist." In Tysiachniouk and McCarthy, eds. *Towards a Sustainable Future: Environmental Activism in Russia and the United States*, 1–18. Saint Petersburg: Saint Petersburg State University.

Tysiachniouk, Maria, and Jonathan Reisman. 2004. "Co-managing the Taiga: Russian Forests and the Challenge of International Environmentalism." In Ari Aukusti Lehtinen, Jakob Donner-Amnell, and Bjornar Saether, eds., *Politics of Forests: Northern Forest–Industrial Regimes in the Age of Globalization*, 157–175. Aldershot, UK: Ashgate.

United Nations (UN). 1982. *World Charter for Nature*. http://www.un.org/documents/ga/res/37/a37r007.htm/.

United Nations Development Program Russia (UNDPR). 2005. *Russia in 2015: Development Goals and Policy Priorities*. Moscow.

United Nations Division for Sustainable Development (UNDSD). 2002. *Russian Federation: Country Profile.* http://www.un.org/esa/agenda21/natlinfo/countr/russia/index.htm/.

Weiner, Douglas R. 1988. *Models of Nature: Ecology, Conservation, and Cultural Revolution in Soviet Russia.* Bloomington: Indiana University Press.

Weiner, Douglas R. 1999. *A Little Corner of Freedom: Russian Nature Protection from Stalin to Gorbachev.* Berkeley: University of California Press.

World Bank. 2004. *Russian Economic Report,* 7 (February). http://siteresources.worldbank.org/INTENVMAT/641999551162240805462/21127284/7Governanceinthe.pdf/.

World Commission on Environment and Development (WCED). 1987. *Our Common Future.* Oxford: Oxford University Press.

World Wildlife Fund Russia (WWFR). n.d. "Model Forests of Russia." http://www.wwf.ru/pskov/eng/mforest/mforest.htm/.

World Wildlife Fund Russia (WWFR). 2007. *Deistvui Legal'no.* Moscow: Forest Program, July. http://www.wwf.ru/resources/publ/book/235/.

Yablokov, Aleksei. 1998. "The Draft Concept of Environmental Policy for the Russian Federation." *Towards a Sustainable Russia* 3 (7): 5.

Yanitsky, Oleg N. 2000. "Sustainability and Risk: The Case of Russia." *Innovation: The European Journal of Social Science Research* 13 (3): 265–277.

Zakharov, Vladimir M. 1999. "First Congress, Second Congress...." *Towards a Sustainable Russia* 1 (12): 6.

Zubarevich, Natalya. 2003. "Russia Case Study on Human Development Progress toward the MDGs at the Sub-national Level." Occasional Paper, Human Development Report Office, United Nations Development Program, New York.

3

Places and Identities on Sakhalin Island: Situating the Emerging Movements for "Sustainable Sakhalin"

Jessica K. Graybill

Socioeconomic and environmental change on Sakhalin Island in the post-Soviet era is largely occurring due to multinational-led offshore hydrocarbon development in the Sea of Okhotsk. Questions about the local socioeconomic benefits and environmental soundness of such development are being raised at all levels, from local to international. Part of the change associated with the economic transformation is an emerging focus on environmental justice and sustainability issues in this region. Concerns are being raised by different sets of actors with different interests in—and different visions for—creating "sustainable Sakhalin." Multiple actors with different visions of Sakhalin's future are shaping its socioeconomic, cultural, and environmental landscapes. Thus a suite of roles and practices of engagement with sustainability issues is developing among different communities on the island. This is reflected in the emergence of multiple new sociocultural and environmental movements on post-Soviet Sakhalin. Although often discussed in terms typical of the larger environmental justice movement (e.g., environmental degradation, environmental racism, environmental justice as social justice; see Benford 2005; Pellow and Brulle 2005), Sakhalin's environmental justice movements do not always overlap and are sometimes contested among the actors involved.

To understand the various strategies employed to create environmental justice on Sakhalin, this chapter addresses the dominant discourses about sustainability among the many actors involved in the island's socioeconomic, political, cultural, and environmental transformation. Focusing on these discourses reveals how dominant actors on Sakhalin perceive sustainability and how the island's different communities perform dialogues about it within and between themselves. It also reveals as much about shifting identity politics and social movements as it does about ideas of sustainability.

Important to this analysis of perceptions of sustainability are theories about the formation of socio-environmental movements and environmental identities. Many scholars and activists recognize that working toward socio-environmental justice involves networks of actors from the local to the global level working together to achieve positive changes in local environments (Castells 1997; Cable et al. 2005).[1] Identifying emerging social movements provides a starting point for understanding how networks of actors within social movements create specific, and perhaps new, identities. Such analysis is useful as well for understanding how socio-ecological movements are defined and navigated by individual actors and communities on Sakhalin.

Acknowledging that environmental justice movements in the twenty-first century have "matured and diversified" and that they are no longer always locally initiated or sustained and are instead imbued with an "international worldview" (Bullard 2005) is my starting point for analyzing actor and identity formation on Sakhalin today. Environmental justice issues arise and are contested precisely because differing worldviews create different understandings of sustainability and of the roles of different actors (e.g., expatriates associated with multinational corporations, government officials, the island's citizens) in creating "sustainable Sakhalin."

Sakhalin's internationalism and the international attention it has garnered make it unique in the post-Soviet era. Oil and gas exploration and extraction and the building of new island-wide production infrastructure by multinational hydrocarbon corporations have incorporated Sakhalin into the global economy. The rapid pace of the island's "globalization" commands the attention of multiple actors: policy makers, multinational corporations, activists, scholars, and citizens. Because Sakhalin's local socio-environmental struggles are internationally produced, supported, and sustained, it is important to critically define the networks of interactions among the socio-environmental actors to understand how environmental justice movements are formulated there (Pellow and Brulle 2005; Agrawal 2006). In other words, what are the "politics of signification" (Hall 1982) on Sakhalin Island today and how are movements for socio-environmental justice and sustainability related to them?

Increasingly important to environmental justice movements on Sakhalin is the role of governance over both resources and the environment. Thus inquiring about Sakhalin's "environmentality" (environmental governmentality; Agrawal 2006), a concept that connects environmental regulation with the agency and practices of individual actors, is also pertinent. Because environmentality seeks to understand how local people begin to care for the environment from the bottom up, it relates to the formation and expression of environmental identities by individual actors and communities alike. The following questions are important to both environmental justice and environmentality: What are the environmental politics that shape the formulation of environmental knowledge? How is power articulated in these politics? How are social practices regulated by existing institutions, and does regulation change any existing behaviors?

This chapter aims to shed light on the ways different actors with emerging, distinct environmental identities engage in a variety of discourses about "Sustainable Sakhalin." Only by exposing the architecture of the "Sustainable Sakhalin" movements in their local and global contexts will we know where they might go in the future. Focusing on the structure of the movements provides (1) background for understanding the different worldviews of people on this island and the components they deem necessary for creating better futures, and (2) a cognitive map of who is shaping what visions of Sakhalin's future for what and whose purposes. By exploring place-based identities and their roles in creating local discourses about sustainability on Sakhalin, this chapter illuminates the island's sustainability landscape and poses provocative questions about the ability to achieve a "sustainable Sakhalin."

Placing Post-Soviet Sakhalin

A brief look at the island's geographic and historical context is helpful in understanding current movements for environmental justice and sustainability on Sakhalin (see Anthony 2005). Located in the Sea of Okhotsk off Russia's far eastern coast, Sakhalin Island has long been called the "edge of the world" (*krai mira*) by local inhabitants and visitors (figure 3.1). For centuries, outsiders have laid claim to the island: first the Mongols and Chinese, then the Japanese, and most recently the Russians, who, having settled on Sakhalin in 1853, officially took possession in 1875, when the Japanese ceded the island to them (Stephan 1971; Vysokov 1996).

Sakhalin was retaken by Japan during the Russo-Japanese War of 1905, then divided between the two countries at war's end, with the

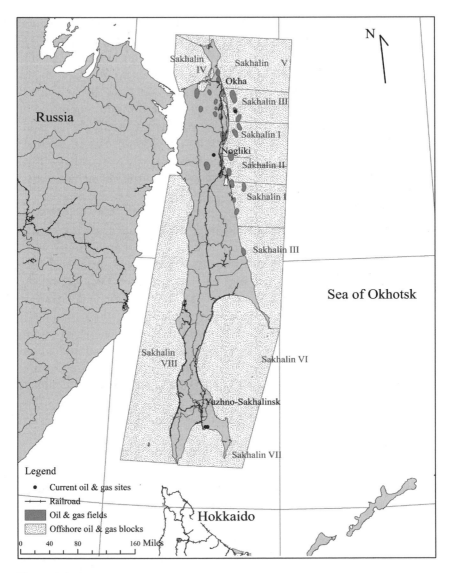

Figure 3.1
Map of Sakhalin Island, showing offshore oil and gas blocks and study locations
(Okha, Nogliki, Yuzhno-Sakhalinsk).

northern half (above latitude 50° north) going to Russia (later the Soviet Union) and the southern half to Japan. The Soviets seized control of the entire island in 1945. In the course of Sovietizing Sakhalin, entire Japanese villages and towns were razed and Japanese settlers forced to return to Japan.[2] The regional capital, Yuzhno-Sakhalinsk, was established at this time on the southern half of the island.

Sakhalin has long served as a resource periphery for multiple countries, that is, a place where renewable or nonrenewable resources are extracted, processed (often only to a limited degree), shipped away, and sold elsewhere. For well over a century, the central governments of the Russian Empire, the Soviet Union, and now the Russian Federation have treated Sakhalin as an "edge" of the world, from which source materials were to be extracted for state benefit. Indeed, it was to acquire and maintain rights to extract oil and marine resources that Japan pushed for control over Sakhalin. The historical treatment of the island's territory and people as a resource periphery has created an economy and culture in which the extraction of forest, fish, coal, and oil products is understood to be Sakhalin's only viable option. According to one local resident, an extractive relationship with Sakhalin's environment is "only natural in such a wild place." For most of the twentieth century, industries based on the extraction of forest, fish, and onshore oil resources (the last in northern Sakhalin) dominated local production (Stephan 1971; Wood and French 1989; Vysokov 1996).

Today, it is offshore extraction of oil and natural gas that drives the island's resource economy. As a result of intensive long-term investment in its production infrastructure, Sakhalin's "production lifetime" has been extended until at least 2035 (Minakir and Freeze 1994; Thornton and Ziegler 2002). Sakhalin is unique in Russia's oil and gas landscape because the technology and capital investment needed to extract its offshore reserves in the Sea of Okhotsk have been provided by multinational joint-venture companies (namely, Sakhalin Energy, ExxonNeftegas, British Petroleum-Rosneft) through production-sharing agreements (PSAs) between the Russian government and foreign partners in the Sakhalin-1 (ExxonNeftegas) and Sakhalin-2 (Shell, Mitsui, Mitsubishi, operating as Sakhalin Energy) projects. Written in 1994, Sakhalin's PSAs were adopted at a time when Russia was desperate to attract foreign investment and when investors required protection of large-scale investments through legally binding international agreements. Sakhalin-1's

PSA is less problematic for Russia than Sakhalin-2's because the federal government already receives profits from Sakhalin-1.

Under Sakhalin-2's production-sharing agreement, however, the Russian government will profit from hydrocarbon development only after most production costs have been recouped by investors. Because Sakhalin Energy's projected costs have more than doubled since the first estimate—to at least $22 billion—the Russian government may have quite some time to wait.

Another unique aspect of Sakhalin-2 is that, until December 2006, it was the only large-scale energy project in Russia operating without a Russian partner. In late 2006, however, Russia's federal and regional governments voiced their objections to the "too little too late" asset distribution scheme in the Sakhalin-2 production-sharing agreement (Rutledge 2004). After expressing its concerns about environmental degradation—in particular, threats to the Grey Whale population in the Sea of Okhotsk and to salmon spawning grounds in Sakhalin's rivers—and about environmental injustice on the island due to Sakhalin Energy's activities (Bradshaw 2006; Elder 2006), the federal government issued an injunction against the company on September 18, 2006, forbidding work on the Sakhalin-2 project. To have the injunction lifted and resume work under a regime of "stability" (Kramer 2006), Sakhalin Energy ceded majority interest in Sakhalin-2 (50 percent plus one share) to Gazprom, Russia's state-owned gas company, on December 21, 2006.

Thus, making full use of environmental rhetoric, the Russian government was able to wrest control of Sakhalin-2 from foreign hands and firmly secure state participation in Sakhalin's energy development. This trend is not unique to Sakhalin: other foreign investors in Russia's hydrocarbon reserves are also being asked to renegotiate deals made in the 1990s (e.g., BP-TNK's investment in the Kovykta Field near Irkutsk; Boykevich 2006).

Ultimately, this struggle over control of Sakhalin's hydrocarbon resources is most detrimental to local environments and the people who depend on them for their survival. With the transfer of majority interest in Sakhalin-2 to Gazprom, financing institutions such as the European Bank for Reconstruction and Development (EBRD) will no longer consider finance packages for the project (Williams 2007). Although the EBRD had previously informed Sakhalin Energy that they found the company's environmental impact assessment "not fit for purpose" (May and October 2005), they remained willing to consider funding the project once environmental issues had been resolved. However imperfect, this "checks and balances" approach (along with international environmental and indigenous activism) worked to help ensure that Sakhalin's hydrocarbon development would be conducted according to international standards for the island's environment and people. Sakhalin Energy's diminished and Gazprom's majority control over development is not likely to result in the greening of Sakhalin-2's development, the first stage of which had already taken place before December 2006, when Sakhalin Energy was the major operator (Bradshaw 2007). Although Sakhalin Energy remains in a managerial position for the project, it is doubtful whether further greening will take place with Gazprom as majority shareholder.

World Wildlife Fund Russia (2007) has requested dialogue with Gazprom regarding Sakhalin-2's compliance with international environmental and social standards but has, as yet, received no direct reply. It remains to be seen whether further greening or even any browning of this project will occur.

In late 2006, the strategy for many actors concerned about the environmental and social well-being of people and place on Sakhalin changed: instead of fighting multinational hydrocarbon corporations, actors formerly siding with the government against Sakhalin Energy now find their target to be global project financiers, potentially a more difficult target. Given this change in strategy and Gazprom's entry into Sakhalin, international development standards and environmentalists may have a shakier—and the environment a "dirtier"—future here. Indeed, NGOs across Russia are facing increased scrutiny since 2007, and some have been closed for failure to comply with bureaucratic regulations seen by many as both unnecessary and burdensome (Osadchuk 2007).

Multinational joint-venture development of Sakhalin's oil and gas resources is important to environmental justice and sustainability issues on the island for two related reasons. First, it means that the international community is now monitoring how development in this peripheral region of Russia, which is difficult and expensive to access even for most Russian citizens. Until December 2006, Sakhalin Energy sought international backing and financing for the project, which required it to adhere to international norms and standards for production, and which placed Sakhalin-2 in the international spotlight. As one international environmental nongovernmental organization put it, "the world is watching"

Sakhalin's unfolding development (http://www.pacificenvironment.org). This increased attention provided hope that international oil and gas development on Sakhalin would be more environmentally friendly than Russian-only development. Second, multinational joint-venture development also means that, through international monitoring and funding, the concerns of local citizens and indigenous peoples are being taken into account in the development scheme (by, for example, creating local jobs, employing local peoples in the projects, and providing new infrastructure). For the first time, local peoples are asked what they want for their futures and are considered part of the development process, not simply recipients of an economic plan imposed on them. On the participation of indigenous people in the development process, Sakhalin Energy wrote in 2007 (with the blessing of some indigenous groups on Sakhalin) that "[we] have been working extensively with the Island's indigenous communities and their representative council to jointly develop an indigenous peoples development plan. The Company has now entered into a partnership with these communities in what is acknowledged as being a truly participatory and transparent process" (http://www.sakhalinenergy .com).

Although the economic transformation of Sakhalin Island through hydrocarbon development seems promising for Sakhalin Oblast, and the potential for long-term socioeconomic improvement is huge, actual socioeconomic benefits (stable jobs, dividends, and new infrastructure for the island's people and its government) continue to be lacking, or do not live up to what has been promised local citizens (Meier 2000; Wilson 2000; Wilson 2003; Bradshaw 2005). Multinational corporations are highly visible in the new island-wide infrastructure for resource development (roads, pipelines, storage facilities, liquefied natural gas plant) and in the new guarded expatriate living complexes ("American Village" and "Strawberry Hills") of Yuzhno-Sakhalinsk. Yet these new features on Sakhalin's landscape exist alongside old crumbling buildings that lack heat, electricity, and telephones (Meier 2000) and leaking onshore pipelines from the Soviet era.

The international community has questioned the efficacy of laws and regulations adopted to protect the island's environment (especially, Sakhalin's salmon spawning grounds and the Grey Whale population in the Sea of Okhotsk; see Newell 2004; EBRD 2007; http://www.sakhalin .environment.ru/ and http://www.pacificenvironment.org) and the rights of indigenous people to land and property (Pika and Grant 1999; Wilson

2003). As recent news headlines about Sakhalin suggest, environmentalism is emerging as a powerful new mode of expressing discontent and demanding change. Significantly, the rhetoric of socio-environmental justice sustainability is important to a range of actors on post-Soviet Sakhalin Island, from government leaders to the multinational corporations to local people. As the world watches the unfolding of a new kind of energy development in post-Soviet Russia, the concept of "Sustainable Sakhalin" is most vocally employed by the multinational corporations associated with the socially and environmentally responsible offshore development of Sakhalin's oil and gas reserves. However, the concept of sustainability has caught the attention of other actors in this region, including environmental NGOs (ENGOs) and indigenous peoples, who employ it in their campaigns to promote the protection of human and natural communities in the face of globalization and long-term hydrocarbon development.

What do environmental justice and sustainability rhetoric, now commonplace in this rapidly globalizing region of Russia, mean to the various people who affect, and are affected by, Sakhalin's hydrocarbon development? What are their differing perceptions of sustainability and sustainable development on Sakhalin? Are these differing perceptions leading to the emergence of multiple movements toward "sustainable Sakhalin"?

Mapping Visions of Sustainability

Considering perceptions of sustainability by actors on Sakhalin allowed me to investigate their different understandings of sustainable (*ustoichivoe*) or responsible (*otvestvennoe*) development in general, and of sustainability on Sakhalin in particular.[3] Investigating perceptions gets at the sources, or epistemological foundations, of conflict. Divergent knowledge systems (perceptions) are as much a source of conflict as divergent values are (Kearney and Kaplan 1997), and investigating actors' perspectives illuminates how people "know" a topic (Dolšak and Ostrom 2003). Taking this one step further, knowing the kaleidoscope of perspectives on sustainability allows one to build the foundation for collaborative, effective, and sustainable environmental decision making (Berkes and Folke 2000; World Bank 2003).

To elicit such perceptions from multiple local actors in sites of hydrocarbon development across Sakhalin (Yuzhno-Sakhalinsk, Nogliki,

Okha), I used conceptual content cognitive mapping (3CM). Anne Kearney and Stephen Kaplan (1997) have found the 3CM method to be a viable way for illuminating the "hypothesized knowledge structures embodying people's assumptions, beliefs, facts, and misconceptions about the world." Others have identified cognitive maps as "guides to the possible" and as ideological constructs (Brody 1981; Monmonier 1993). To all this, I would add that, on Sakhalin, cognitive maps act as spaces for understanding the (dis)connections between different actors and allow the possibility for new dialogue about a sustainable Sakhalin to develop.

Actor-participants in this study identified themselves as local or indigenous people, local environmental professionals, professional environmentalists, government leaders, or Western expatriates. "Indigenous people" are Nivkh or Evenk. "Local people" are people who have lived on Sakhalin for ten or more years. "Local environmental professionals" (hereafter called "local professionals") are distinguished from local people in that they work in environmental capacities, whether academic, public, or private (e.g., as environmental engineers, consultants). "Government leaders" may be indigenous or Slavic, but identify themselves, first and foremost, as government representatives. "Professional environmentalists" are local or international people working for ENGOs whose main concern is environmental protection. "Western expatriates" (hereafter called "expatriates") are businesspeople from Europe and North America.

In separate tasks, all actor-participants wrote down and subsequently organized their perceptions of (1) sustainable or responsible development in general and (2) sustainability on Sakhalin in particular. After they completed each of these tasks, they also provided narratives for the cognitive maps, often adding in additional comments or thoughts about sustainability in this region. The names of research participants, hereafter referred to in the third person and by their work affiliations, have been kept confidential to maintain anonymity.[4] I conducted interviews in Russian with Russian, Nivkh, and Evenk participants and in English with expatriate participants.

Sustainability Discourses

From the cognitive mapping and interview data gathered during seven months in 2005 while living among different actors on Sakhalin, I identified five emerging, yet dominant discourses about sustainability. They are not always clearly stratified according to different types of actors: the discourses reach across the actor types. In addition, discourses about sustainable or responsible development are intertwined with notions of identity and reason for being. To highlight the interplay between them, these notions are placed next to the five dominant discourses in table 3.1.

Discourse 1: Sustainability as Diversity
Local people, local professionals, and expatriates identified sustainability as inherently incorporating diversity into its definition. Although diversity holds multiple meanings among these participants, in terms of Sakhalin's economic future, diversity was thought by most to be important in the types of (1) industry and associated jobs and (2) education and training to obtain those jobs available to Sakhalinites on the island.

First, economic diversity is commonly identified by expatriates and local people alike as an underlying key to Sakhalin's sustainability. One expatriate said, "Energy development [should be treated] as a pillar, as one piece of a foundation to building or developing other things. It is an engine [for future development], not heaven." Another expatriate, perceiving the lack of other types of large-scale industries (e.g., agriculture) on Sakhalin to be problematic for a self-sustaining future for Sakhalin ("No oil and gas development is sustainable forever"), asked rhetorically, "Where's the beef?"

Herein lies the catch. From a local person's perspective, jobs that pay well and that have futures currently do not exist outside the oil and gas industry. Indeed, one local professional noticed a mini brain drain on Sakhalin, as workers were pulled from potentially viable industries and job sectors into the oil and gas sector *at all levels* (from unskilled labor to secretarial to managerial) precisely because of the dominance of the hydrocarbon industry. They noted that well-qualified workers who formerly filled roles in government institutions (e.g., bookkeepers, educators, health care providers) are recruited away by the oil companies, a process that halts the development of other sectors and undermines the competence of government institutions to deal with economic transformation.

Second, diversity in education and training was understood as necessary for developing a sustainable Sakhalin. Many participants recognized that Soviet-era and even post-Soviet Russian education and training for jobs are inadequate for dealing with Sakhalin's globalizing economy

Table 3.1
Differences and similarities among actors in sustainability discourses and perceptions of identity on Sakhalin Island

	Indigenous people	Local people	Local professionals	Local government	Environmentalists	Expatriates
Sustainability discourses						
Discourse 1: Sustainability as diversity		X	X			X
Discourse 2: Sustainability as social change	X			X		X
Discourse 3: Sustainability as a healthy environment	X	X	X	X	X	
Discourse 4: Sustainability as balance	X	X	X		X	X
Discourse 5: Sustainability as leading industry's responsibility	X			X		
Identity perceptions						
Sakhalin as homeland	X	X				
Sakhalin as maldeveloped						
Prior to presence of multinationals	X		X		X	
During presence of multinationals				X		X
Sakhalin as beautiful wilderness		X	X		X	
Salmon as survival	X	X			X	X
Oil as death (of nature, of humans)	X	X		X	X	
Oil as only future for Sakhalin	X		X	X		X

and with multinational employers on the island; new kinds of education and training are needed. For example, specialists (e.g., bookkeepers, economists) have had to retrain, often at their own expense, for transitional and market economic systems because their education and training were seen as inadequate by multinational corporations. As a result, local people question the value of advanced degrees in the transition era, and many give up and instead find jobs in the burgeoning and lucrative service industry (e.g., tourism, hotel and restaurant service).

Although this problem is recognized by many actors, it is hard, if not impossible, to provide diversity through local schooling linked to a national education system in need of serious updating (Mindeli and Pipiya 2002). In the meantime, those who can afford privatized universities and technical colleges get the training they need to operate in a globalizing economy, leaving the majority behind, unable to compete. This situation will have to change, one indigenous Sakhalinite strongly felt, if the Sakhalin community is ever to end its dependence on government or multinational handouts and on unsustainable resource use (e.g., overharvesting fish and plants for subsistence from season to season)—and to become truly sustainable.

Discourse 2: Sustainability as Social Change

Indigenous people, local government leaders, and expatriates recognize sustainability as predicated on social change and adaptation to new circumstances (e.g., a market economy, global actors, changing relationships to the outside world). An insular place in both tsarist and Soviet eras, Sakhalin was a penal colony under the tsars and a closed military and border zone under the Soviets, who prohibited even Russian citizens from traveling there without proper paperwork. Sakhalinites often regard themselves as markedly different from Russians of the mainland (*materik*), and pride themselves on their ability to survive through subsistence and community connections.

Noting this insularity in business dealings, one expatriate viewed it as needing to change: "Fortress Sakhalin will not lead to sustainable Sakhalin." "Oil and gas investment affects everything here," he went on to say. "It is not an opportunity that people can choose to operate in: it is forcing people to operate in it to survive." He envisioned a sustainable Sakhalin where local people, not just activists, begin to engage with the changes to local settings (economy, culture, environment) in the post-Soviet period. "Russians," he felt, "need to take responsibility for their own development."

A local governmental official echoed this sentiment, saying, "[Local] people can't just go into their apartments and shut the door to forget the world outside. We have to look around and start thinking about a wider world." Some people are beginning to do exactly that: one indigenous actor noted that "these projects push people to think about their 'place.' ... [I ask myself about] my place: me and nature, me and society, me and the government, me and the multinational companies."

Interestingly, both indigenous and expatriate actors feel that in order for sustainability to happen on Sakhalin, there would need to be a change in the mentality of local *Russian* (Slavic) people. One expatriate noted that "Russian culture and mentality limits development. Sakhalin-ites are not worldly. They are still island people and they preserve that identity." Another expatriate perceived a "resistance to improvement," or progress, through multinational development: "If I were a cartoonist, I would draw oil companies digging in and trudging forward. Russians would also dig their heels in and pull backward, but would be yelling, 'Hurry up!'"

Yet another expatriate noted a resistance to changing socioeconomic circumstances that was problematic for any kind of development: "Sakhalinites need to become open to ideas, information, and other social models, morals, etc. [They] need this diversity for sustainability to happen." In this sense, "locals" should become socially and culturally more sensitive to the wants and needs of "globals."

Several indigenous actors also felt it was local Russian citizens who needed to change the most. Nivkh and Evenk dislike how Russian citizens on Sakhalin often consider their own plight to be the same as that of indigenous peoples. (One Russian citizen that I met on Sakhalin categorized all indigenous people as "the same as us" yet also "lazy and spoiled" due to the government subsidies allotted in the Soviet past.) This openly spoken sentiment does not help Russian-indigenous relations. Nivkh and Evenk with whom I spoke felt themselves to be different in many ways, but what came across most often was the sense of being different physiologically (craving fish or caribou meat, unable to eat certain Russian foods, intolerance for alcohol, being attuned to a seasonal clock rather than a daily clock, etc.). These actors feel strongly that Russian citizens need to learn to respect and accommodate the island's native human diversity in order for all parties to work toward a sustainable future. Asked whether any wedges had been driven between indigenous and local Russian citizens due to differences over indigenous activismfor social change in the transition era, one indigenous person replied, "No," but after a pause added, "Not yet."

I include these explanations and quotations to highlight the tensions often present in interactions between different actors on Sakhalin due to historical and current factors and (mis)readings of culture and society.

Discourse 3: Sustainability as a Healthy Environment

All actors except expatriates identified sustainability as a healthy environment. "Healthy" was understood to mean cared for, unpolluted, and (often) pristine. "Environment" was understood to include offshore marine environments in the Sea of Okhotsk as well as all terrestrial and aquatic environments on Sakhalin Island. This concept of environmental health is reflected in identifying Sakhalin as a homeland, salmon as a means of survival, and oil as "death" by indigenous peoples and local peoples who largely depend on Sakhalin's environmental resources for daily survival in the transition era.

Notably, the only category of actors that did not relate sustainability to healthy environments was the expatriates. When asked about the role of the environment in developing sustainable Sakhalin, one expatriate said that "I am not as worried about the environment as I am about social issues. The footprint of the oil and gas industry is less than from other industries. I have undue faith in nature to take care of herself." Another explained that the "oil and gas impact [on Sakhalin] is not so big because here the problem is more about the terrible heritage of waste and treating nature poorly. That already exists here." In both of these quotations, there is a sense of "Why bother to create a healthy natural environment on Sakhalin when other industries are more polluting and the existing environment is already unhealthy?"

Discourse 4: Sustainability as Balance

All categories of actors except local and regional government leaders identified balance as a necessary component of sustainability. Balance largely means the need to develop equal distribution of profits and benefits to different actors across Sakhalin, Russia, and abroad. Unbalanced benefits include job creation for some people but not for others, higher salaries for foreigners, and disproportionate profit taking by actors in Moscow and abroad. Among many local people without hydrocarbon-related jobs, there is resentment over Sakhalin's growing role as a resource periphery and toward the growing elite "oil class" on the island, whose members profit from the offshore projects while others do not.

Most actors feel that creating sustainability through balanced benefits depends on strong local and regional governments that are interested in creating benefits for Sakhalinites, not only for politicians and government bureaucrats. An international businessperson believed it necessary "to make regional and local governments responsible for their people" on Sakhalin. A local professional from Nogliki believed this lack of leadership to be responsible for economic imbalance. The professional further believed that individual regions, such as Nogliki Raion or Sakhalin Oblast, have "weak control over our federal government" because local people should have the power to demand that cement (for new pipeline infrastructure) be made on Sakhalin and not imported from Japan (as it currently is). "Investors don't care where cement [for the projects] comes from, so for them it's fine if it comes from Japan and not Nogliki. This has to get under control."

Local people largely attribute this imbalance in job creation, purchasing, and industry development to investors behind the multinational companies who drive the development decision-making processes. This is expressed lucidly by a international businessperson, who claimed that the (im)balance "is all about turf battles. While they fight, Rome is burning. Sakhalin government needs to provide strategic leadership." "They," here, are the local and regional governments, which have different interests in and potential gains from the projects, as different administrative organs have different governance directives and structures. Thus the governments should begin to coordinate their efforts to manage multinational companies instead of continuing to argue among themselves.

Different social groups in this formerly Communist region are beginning to understand the imbalance between profits from hydrocarbon development and interests in creating a sustainable Sakhalin as social injustice. Many local professionals blame the new social stratum—the oil class: "The island is divided into those who serve the oil and gas industry and those who don't." A younger member of this class, having been trained at the privatized economics institute, noted that "we are the golden children [*zolotye deti*] of Sakhalin. We will thrive because of the oil opportunities here." On the other hand, older local professionals worry that such uneven benefits from Sakhalin's transition to a market economy will change local values from Communist to capitalist or materialist. Local government leaders and environmentalists identify this with current problematic development of the island.

Discourse 5: Sustainability as Leading Industry's Responsibility

Two categories of actors, indigenous people and local government leaders, strongly feel that creating sustainable Sakhalin is the task of the multinational oil and gas industry, the island's leading industry. This view of sustainability's ideal source is strongly related to past understandings of the roles of industry and government. In the Soviet era, all industry was owned by the government, which was responsible for taking care of its citizens "from cradle to grave" under Communist doctrine. The collapse of government-run industry and the end of government subsidies has made a particularly bitter impression on those who depended the most on them. In the case of Sakhalin, this was the local and regional governments (which depended on subsidies to continue operating in the economically unviable Far East and which must now fund their own budgets without assistance from the federal government) and indigenous peoples (who depended on subsidies to travel, to obtain higher education, and even to rent their apartments or pay utility bills; see Bartels and Bartels 1995; Grant 1995; Kerttula 2000).

More than any other discourse about sustainability, discourse 5 reflects mourning for the Soviet past and lack of confidence in a positive post-Soviet future on the part of certain categories of Sakhalinites. Thus both indigenous people and local government leaders identify Sakhalin as "maldeveloped" in the post-Soviet—but *not* in the Soviet—period.

Sustainability, Identity, Action: Emerging Socio-Environmental Movements

Sakhalin is not alone in experiencing changes to its economy, society, and environment in the post-Soviet era. Like many other communities across the former Soviet Union (FSU) in the early transition era, local peoples have been forced to turn back, at least in part, to a subsistence mode of living, a process known as "ruralization." Also like many other FSU communities, Sakhalinites are beginning to raise questions about their cultural identities, their future, and their rights to a healthy environment. Forced socioeconomic transformation through multinational hydrocarbon development on this formerly insular island is changing how people value and experience their environment. Environmental degradation wrought by such development directly threatens the survival of many people on Sakhalin, particularly those located outside of the regional capital, Yuzhno-Sakhalinsk.

Combinations and networks of indigenous, local, and transnational actors and interests on Sakhalin today are transforming the island's environmental and sociocultural landscapes for all its residents. Both local and international forces have led to the emergence of environmental justice movements on Sakhalin emphasizing the protection of marine and riverine environments and indigenous peoples' rights to land and resources.

Weaving together the sustainability discourses and perceptions of identity on Sakhalin provides a background for understanding the emerging characteristics of Sakhalin's eco-political movements, to which social and environmental justice concerns are central (see table 3.2). Components and actors in these movements will necessarily shift over time, as socio-ecological movements on Sakhalin continue to evolve, but charting today's landscapes provides salient examples of the "distinct voices of the movement [that] can be heard through the cacophony" (Castells 1997, 113).

Distinct environmental identities related to cultural nationalism, environmental justice, radical environmentalism, Western corporate sustainability, and enforceable environmentalism are emerging with their distinct representatives on Sakhalin in the post-Soviet era. Although there is some hybridity and overlap among representatives of these environmental identities, differences in strategy and membership also stand out. The catalysts for much of the movements' activities are largely international ENGOs (specifically, Pacific Environment, Wild Salmon Center, Friends of the Earth) that are focused on creating an environmental and indigenous activism base on Sakhalin. Sherry Cable and colleagues (2005) identify professional environmentalists as actors who are largely concerned with conservation and environmental protection issues, and who target policy makers and the general public to create reform and enforce environmental protection. In Sakhalin's case, professional environmentalists from abroad have joined local organizations to coordinate sophisticated international environmental and indigenous activism aimed at (1) creating international awareness of socio-environmental (in)justice on Sakhalin and (2) halting development that is not environmentally or socially sound.

Table 3.2
Emerging identities, movements, and strategies on Sakhalin Island

Movement identity	Actor identity	Representatives (organizations)	Movement strategies
Cultural nationalism	Indigenous peoples	Sakhalin Indigenous Minorities Council Kik-kik Russian Association of Indigenous Peoples of the North (RAIPON)	Demonstrations/protests, demanding ethnological impact assessments (*etnologicheskie ekspertizy*), creating new "traditional" holidays such as Day of the Caribou Herders (Den' Olenevodov), forming social movement organizations, traveling to and participating in international indigenous forums, encouraging/supporting academic research by anthropologists
Environmental justice	Local-regional government Local-global environmentalists	Sakhalin Environmental Watch Pacific Environment Wild Salmon Center	Local-international demonstrations, appeals to Russian government to halt Sakh-2 activities harmful to indigenous cultures, to local environments and species (salmon and gray whale), or to both
Radical environmentalism	Local-global environmentalists Indigenous peoples Local peoples	Sakhalin Environmental Watch Local peoples Indigenous peoples	Direct action to halt hydrocarbon infrastructure development on the island in the north (roads, pipelines near Nogliki) and in the south (liquefied natural gas plant near Priogorodnoye)
Western corporate sustainability	Multinational and joint-venture hydrocarbon companies	Sakhalin Energy BP-TNK-Sakhalin Rosneft-ExxonMobil	Appealing to local communities to participate as "equal stakeholders" in surveys and development forums
Enforceable environmentalism	Russian government Local-global environmentalists	Local, regional, and federal governments	Requiring environmental impact assessments (EIAs) before permitting development, demanding environmental or ethnological assessments of the impact of hydrocarbon development on indigenous cultures and lifestyles

Figure 3.2
Indigenous protest near Nogliki, Sakhalin Island, against Sakhalin Energy, January 2005.

An example of the sophisticated and international nature of these ENGOs' efforts to date is the coordination of simultaneous protests in Nogliki, Yuzhno-Sakhalinsk, Moscow, London, and New York in 2005 against development of Sakhalin 2 (figure 3.2). Indeed, one might even consider the EBRD's hesitancy to fund Sakhalin-2 before Gazprom's entry into the project as due in part to this international activism. Understanding that "the age of globalization is also the age of nationalist resurgence" (Castells 1997, 27), Sakhalin's movements are operating internationally to create change in the local social order.

Although these activities may not seem new or impressive to those accustomed to protests against socio-environmental injustice worldwide, especially in an increasingly globalizing world economy (consider, for example, the 1999 WTO protests in Seattle), it is new to Russia since 1991. Public expression of disagreement with the status quo was not tolerated in the Soviet period, and the seemingly well-supported presence of both environmental and indigenous activism, and the concrete changes effected by these activists in the island's post-Soviet era development challenge the common perception that Russian civil society is under-developed and will take generations to develop. One expatriate admired the emergent environmentalism of the Sakhalinites, noting: "They're learning. They're learning how to deal with us [multinational corporations]. They're not there yet, but they're getting better."

A distinction must be made between professional environmental movements and environmental justice movements. As Cable and colleagues (2005, 60) argue, those actors fighting for environmental justice are "citizens associated with community-based organizations formed to oppose the disproportionate exposure to environmental risks endured by the working class, the impoverished, and people of color." Although professional environmentalism has taken hold through international conservation and protection campaigns, it remains to be seen whether locally developed and sustained activism can become as sophisticated and coordinated among local actors on Sakhalin.

With the sale of 50 percent plus one share of Sakhalin-2 to Gazprom, it is unknown what the future—and impact—of internationally led professional environmentalism will be on or for the island. The environmentalists' initial response to the December 2006 change in ownership of Sakhalin-2 was to "withhold judgment" (Pacific Environment 2006) and hope for continued adherence to international development standards that. In 2008, U.S. and UK lenders denied Sakhalin Energy financial support due to concerns about the company's ability to meet funding criteria (including environmental criteria; Youngson 2008). Environmentalists celebrated the news as a "tremendous victory" (Sakhalin Environment 2008), although they are wary of celebrating overmuch lest project financing be found elsewhere in the future.

Robert Benford (2005, 51) notes that, when seeking socio-environmental justice within existing "legislative, judicial and regulatory systems, the status quo will continue to be reproduced." Indeed, one might ask, what justice can be sought, for people or the environment, where laws, regulations, and governance are so malleable? Yet the status quo for multinational hydrocarbon development and for the oblast's citizens remains unknown, especially with the recent changes in management and profit structures from Sakhalin-2.

In *Environmentality*, Arun Agrawal (2006, 201) observes that "the government of nature led to the birth of the environment," an observation that aptly relates to the awakening to environmentalism by many actors in the post-Soviet era. But it is not only environmentalism that is important to understanding the emergence of Sakhalin's post-Soviet socio-ecological landscape; the roles played by government and corporate entities are also critical. The struggle over governance of resources and people may remain the most important issue in this resource periphery, which speaks more to the continuity of Sakhalin's Japanese, prerevolutionary Russian, Soviet, and post-Soviet histories than to the changes wrought by multinational hydrocarbon development in the post-Soviet period.

This exploration of how environmental actors are emerging on Sakhalin elucidates what notions of sustainability they carry with them as they promote a better socio-ecological future for the island. It is important to note that not one person defined sustainability as meeting the needs of the current generation while allowing future generations to meet their own needs as well (see Bruntland Commission 1987). "Sustainable Sakhalin," then, can be understood both as serving a rhetorical need and as expressing the desire by many local and global citizens for a better socio-environmental future on this island. My time among local actors on Sakhalin reveals a contested landscape onto which (E)NGO communities and expatriate multinational corporations are attempting to map a Western concept of sustainability—and how to achieve it. They are doing this through fairly traditional means: demonstrating against socio-environmental injustices, getting involved as stakeholders at development planning stages, promoting development of local economies and cultures by encouraging development of small businesses. This concept of environmentalism is contested, however, by local actors who are searching to define themselves and their reasons for being in the post-Soviet era. The current study reveals the regional character of Sakhalin's sustainable development, shedding light on how different actors in the same locale argue differently for better human and environmental futures.

Notes

1. Although "environment" is used broadly here to include its physical, social, and cultural aspects, I define the term a bit more narrowly as the place where people live, work, play, and subsist, subsistence being an important aspect of daily life on Sakhalin.

2. In addition to the Japanese settlers' forced to return to Japan, some native people of Sakhalin (mostly Ainu) were removed from their native homeland on Sakhalin to Hokkaido by Soviet troops (Stephan 1971).

3. Here I use the term *actor* to mean any individual living on Sakhalin Island (temporarily or long-term) who can affect or who is affected by changes in natural resource uses or the spaces thereof. If an actor-participant was unclear what "sustainable" (*ustoichivoe*) development meant, I explained that it meant socially and environmentally responsible (*otvestvennoe*) development.

4. Participants said anonymity was important, if not essential, to maintain a sense of security in answering questions.

References

Agrawal, A. 2006. *Environmentality*. Durham, NC: Duke University Press.

Anthony, C. 2005. "The Environmental Justice Movement: An Activist's Perspective." In D. N. Pellow and R. J. Brulle, eds., *Power, Justice, and the Environment*, 91–100. Cambridge, MA: MIT Press.

Bartels, D. A., and A. L. Bartels. 1995. *When the North Was Red: Aboriginal Education in Soviet Siberia*. Montreal & Buffalo: McGill-Queen's University Press.

Benford, R. 2005. "The Half-Life of Environmental Justice Frame: Innovation, Diffusion, and Stagnation." In D. N. Pellow and R. J. Brulle, eds., *Power, Justice, and the Environment*, 37–54. Cambridge, MA: MIT Press.

Berkes, F., and C. Folke, eds. 2000. *Linking Social and Ecological Systems*. Cambridge: Cambridge University Press.

Boykevich, S. 2006. "TNK-BP Moves to Solve Kovykta Row." *Moscow Times*, March 15.

Bradshaw, M. J. 2005. "Environmental Groups Campaign against Sakhalin-2 Project Financing." *Pacific Russia Oil and Gas Report* 3: 13–18.

Bradshaw, M. J. 2006: "Russia's Oil and Gas: State Control, the Environment, and Foreign Investment." *World Today* 18–19.

Bradshaw, M. J. 2007. "The 'Greening' of Global Project Financing." *Canadian Geographer* 51 (3): 255–279.

Brody, H. 1981. *Maps and Dreams: Indians and the British Columbia Frontier*. Long Grove, IL: Waveland Press.

Brundtland Commission (World Commission on Environment and Development). 1987. *Our Common Future*. Oxford: Oxford University Press.

Bullard, R. D. 2005. "Environmental Justice in the Twenty-First Century." In *The Quest for Environmental Justice: Human Rights and the Politics of Pollution*, 19–43. San Francisco: Sierra Club Books.

Cable, S., T. Mix, and D. Hastings. 2005. "Mission Impossible? Environmental Justice Activists' Collaborations with Professional Environmentalists and with Academics." In D. N. Pellow and R. J. Brulle, eds., *Power, Justice, and the Environment*, 55–76. Cambridge, MA· MIT Press.

Castells, M. 1997. *The Power of Identity*. Oxford: Blackwell.

Dolšak, N., and E. Ostrom. 2003. *The Commons in the New Millennium: Challenges and Adaptation*. Cambridge, MA: MIT Press.

Elder, M. 2006. "Mitvol Tries to Allay Investors' Fears." *New York Times*, November 15.

European Bank for Reconstruction and Development (EBRD). 2007. "EBRD No Longer Considers Current Financing Package for Sakhalin II." Press release. January 11. http://www.ebrd.com/new/pressrel/2007/070111.htm/ (accessed March 17, 2007).

Grant, B. 1995. *In the Soviet House of Culture: A Century of Perestroikas*. Princeton, NJ: Princeton University Press.

Hall, S. 1982. "The Rediscovery of 'Ideology': Return of the Repressed in Media Studies." In M. Gurevitch, T. Bennett, J. Curran, and J. Woollacott, eds., *Culture, Society and the Media*, 56–90. London: Methuen.

Kearney, A., and S. Kaplan. 1997. "Toward a Methodology for the Measurement of Knowledge Structures of Ordinary People: The Conceptual Content Cognitive Map." *Environment and Behavior* 29 (5): 579–608.

Kerttula, A. M. 2000. *Antler on the Sea: The Yup'ik and Chukchi of the Russian Far East*. Ithaca, NY: Cornell University Press.

Kramer, A. E. 2006. "Gas Investors Bow to Pressure on Recovering Expenses." *New York Times*, December 29, sec. C, p. 7.

Meier, A. 2000. "The Breaking Point: Despite Abundant Natural Resources, the People of Sakhalin Are Just Scraping By." *Time*, October 31.

Minakir, P., and G. L. Freeze. 1994. *The Russian Far East: An Economic Handbook*. Armonk, NY: M.E. Sharpe.

Mindeli, L. E., and L. K. Pipiya. 2002 "How Is Basic Science in Russia to Be Preserved?" *Herald of the Russian Academy of Sciences* 72 (2): 99–106.

Monmonier, M. 1993. *Mapping It Out: Expository Cartography for the Humanities and Social Sciences*. Chicago: University of Chicago Press.

Newell, J. 2004. *The Russian Far East: A Reference Guide for Conservation and Development*. McKinleyville, CA: Daniel and Daniel.

Osadchuk, S. 2007. "NGOs Scramble to Meet Deadline." *Moscow Times*, April 12.

Pacific Environment. 2006. "Sakhalin II Deal Leaves Shell and Gazprom on the Environmental Hook." December 22. http://www.pacificenvironment.org/article.php?id=2150/ (accessed 6 March 2008).

Pellow, D. N., and R. J. Brulle. 2005. "Power, Justice and the Environment: Toward Critical Environmental Justice Studies." In Pellow and Brulle, eds. *Power, Justice, and the Environment*, 1–22. Cambridge, MA: MIT Press.

Pika, A., and Grant, B., eds. 1999. *Neotraditionalism in the Russian North: Indigenous Peoples and the Legacy of Perestroika*. Seattle: Edmonton Canadian Circumpolar Institute.

Rutledge, I. 2004. *The Sakhalin II PSA—A Production "Non-Sharing" Agreement: Analysis of Revenue Distribution*. Prague: CEE Bankwatch Network.

Sakhalin Environmental Watch. 2008. "Sakhalin II Victory—Who is Now Prepared to Touch Beleaguered Project?" March 4. http://www.sakhalin.environment.ru/en/.

Stephan, J. 1971. *Sakhalin: A History*. Oxford: Oxford University Press.

Thornton, J., and T. Ziegler. 2002. *Russia's Far East: A Region at Risk*. Seattle: National Bureau of Asian Research; University of Washington Press.

Vysokov, M. 1996. *A Brief History of Sakhalin and the Kurils*. Yuzhno-Sakhalinsk, Russia: Sakhalin Book Publishing House.

Williams, A. 2007. "EBRD No Longer Considers Current Financing Package for Sakhalin II." http://www.ebrd.com, January 11 (accessed February 12, 2007).

Wilson, E. 2000. "North-Eastern Sakhalin: Local Communities and the Oil Industry." Russian Regional Group Working Paper Series.

Wilson, E. 2003. "Freedom and Loss in a Human Landscape: Multinational Oil Exploitation and Survival of Reindeer Herding in North-Eastern Sakhalin, the Russian Far East." *Sibirica* 3 (1): 21–48.

Wood, A., and R. A. French, eds. 1989. *The Development of Siberia: People and Resources*. London: Basingstoke Macmillan; School of Slavonic and East European Studies, University of London.

The World Bank. 2003. World Development Report 2003: *Sustainable Development in a Dynamic World: Transforming Institutions, Growth and Quality of Life*. Oxford: Oxford University Press.

World Wildlife Fund Russia (WWFR). 2007. "EBRD's Sakhalin Decision Should Be Followed by Remaining Potential Funders, Say Campaigners." Press release. January 15. http://www.wwf.ru/resources/news/article/eng/print/2796/ (accessed March 6, 2008).

Youngson, B. 2008. "Sakhalin Shelves Huge Funding Bid." http://www.energycurrent.com. March 4 (accessed March 6, 2008).

4

Oil Wealth, Environment, and Equity in Azerbaijan

Shannon O'Lear

Azerbaijan became independent in October of 1991, and it was only a few years later, in 1994, that President Heidar Aliyev signed the "Contract of the Century," which opened up the country's oil fields to the international oil industry. The resulting investment of capital and technology allowed Azerbaijan to expand its oil exports—and income generated from those exports—significantly. Despite the surge of oil-generated wealth, however, Azerbaijan's economic development has actually moved backward rather than forward (Rasizade 2003). In many ways, Azerbaijan's oil wealth may be seen as a curse. It has contributed to a growing income gap, increased governmental centralization, and a lack of growth in other economic sectors (O'Lear 2007); there has been limited transparency and oversight in how the country's oil wealth is managed and in the role of foreign companies in its social and economic development (Gulbrandsen and Moe 2007). Within this growing imbalance between the oil industry and other parts of the economy, on the one hand, and between a powerful central government and the general populace, on the other, environmental justice emerges as an issue to consider in this young, oil-rich state.

The environmental justice movement grew out of the merging of environmental with social justice activism. From the outset, environmental justice advocates have focused on the politics behind the disproportionate exposure to environmental hazards experienced by underrepresented or less powerful groups in a variety of contexts. Although much of the initial work recognized correlations between environmental inequity and race or ethnicity, the meaning of environmental justice has expanded to include groups not defined by race or ethnicity, such as women, children, and the poor (Cutter 1995). More recent conceptualizations view environmental justice within a broader spectrum of rights and security

(Agyeman et al. 2003). Whereas affluent and powerful groups of people can move to desirable areas, absorb increased financial costs associated with increasingly scarce resources, and otherwise minimize their exposure to negative impacts of industrialization and globalization, people who are poor, less powerful, or outside the culturally dominant or economically elite groups are often the first to be exposed to negative environmental externalities (Pellow and Brulle 2005). Environmental injustice is experienced locally, but the processes generating it occur at national and transnational levels, involving states and multinational corporations with a global reach.

Although much work has been done to document and study both environmental degradation in the former Soviet realm (Micklin 1992; Peterson 1993; DeBardeleben and Hannigan 1995; Saiko 2001) and the environmental activism it inspired (Yanitsky 1993; O'Lear 1996, 1999), this volume focuses more specifically on environmental injustice in the countries of the former Soviet Union. Previous work on the potential for environmental justice in central and eastern Europe, also part of the former Soviet realm, considered the role of the Communist system in establishing environmental problems and challenges posed by incomplete political and economic transition for developing more equitable environmental conditions (Costi 2003). This chapter considers the Soviet legacy of environmental degradation in oil-rich Azerbaijan and the impacts of that degradation on the country's evolving political and economic systems. Environmental injustice may occur in places where environmental externalities such as industrial pollution and soil degradation are concentrated, putting specific, geographically clustered groups of people at risk. Yet environmental injustice may also be more spatially diffuse: it may include benefits or costs associated with national resource wealth or with access to natural resource–based public utilities (water, electricity, and natural gas). This chapter examines environmental injustice as reflected in a systemic, uneven distribution of the benefits and costs of natural resources, and as perceived by the people affected.

One way to understand linkages between environmental conditions and public perception is through the environmental Kuznets curve, which plots the relationship of environmental quality to economic development. Shaped like a downward-turned U, the curve shows that environmental quality (along the y-axis) first worsens with economic development (along the x-axis), then reaches a turning point, and finally improves as economic development, wealth, and, by extension, capacity to address environmental degradation increases.[1] Echoing Abraham Maslow's widely cited hierarchy of needs, the environmental Kuznets curve implies that the environment will not be a priority until other, basic economic needs have been met. Researchers have applied the concept of the Kuznets curve in a study of citizen complaints about the environment in China (Dasgupta and Wheeler 1997) in order to determine whether public awareness of environmental problems more closely correlated with actual environmental pollution or with income level. The number of complaints about environmental conditions was found to correlate less with the actual occurrence of toxic substances than with income levels. This finding suggests that environmental quality, though an important element of human well-being, is not necessarily a feature of day-to-day life that most people put first or act upon. Which is to say, we cannot assume that poor environmental conditions will necessarily lead to environmental activism (see, for example, Eyerman and Jamison 1991).

As a starting point for assessing environmental justice in Azerbaijan, the chapter looks at the degree to which environmental degradation there coincides with ethnic distribution. A comparison of spatial patterns of environmental degradation and ethnic groups suggests that, in the case of Azerbaijan overall, environmental injustice may not be best described as an ethnic issue. The chapter then touches on the concept of human security as a lens through which to examine impacts of Azerbaijan's oil wealth on conditions of daily life for the country's populace. Human security provides a way to think about environmental conditions as but one dimension of people's well-being. Whether environmental conditions or their health impacts emerge at the forefront of public concerns depends in part on what other concerns compete for priority in the day-to-day lives of citizens. Next, the chapter briefly assesses the impacts of Azerbaijan's oil wealth to provide background on general economic trends in the country. Data from a nationwide survey in Azerbaijan demonstrate that, although people are aware of and have concerns about environmental problems associated with the oil industry, these are eclipsed by other, daily concerns. The survey data let us examine correlations between economic status and environmental concern and between perceived environmentally related health impacts and environmental concern. These survey data illustrate the linkage between economic and environmental conditions—a recurrent theme in recent work on environmental justice.

Ethnic Groups and Environmental Degradation

Because environmental justice is often discussed in ethnic terms (see, for example, Rhodes 2003; Rinquist 2006), it is useful to compare spatial patterns of ethnic concentrations to areas of greatest environmental concern to see whether a correlation exists between them in Azerbaijan. The map in figure 4.1 demonstrates that, although there are three discernible instances of overlap between spatial clusters of ethnic groups and areas of significant environmental concern, none of them directly suggests that any particular ethnic group has been targeted or neglected in terms of environmental conditions. First, in Azerbaijan's northeast-

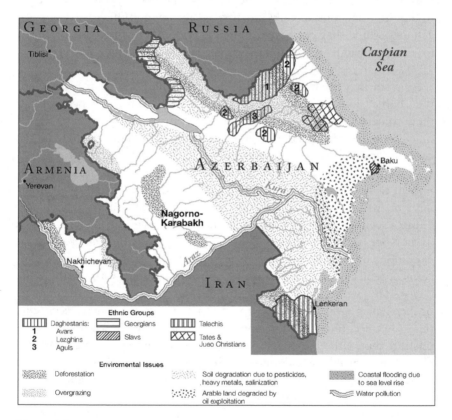

Figure 4.1
Map of ethnic groups and environmental security issues in Azerbaijan (UNEP 2004; http://www.envsec.org/southcauc/ [accessed March 4, 2008]). Used with permission of Philippe Rekacewicz.

ern region near the Russian border, three concentrated subgroups of Daghestanis—Avars, Lezghins, and Aguls—live in areas degraded by overgrazing and deforestation and subject to landslides. Because such environmental conditions extend far beyond the northeastern region, however, Dagestanis are not experiencing more environmental degradation than other ethnic groups elsewhere in Azerbaijan and do not therefore represent a case of ethnically targeted environmental injustice. Similarly, a concentration of ethnic Slavs lives in and around Baku, on the Caspian Sea, an area at risk of contamination from decaying Soviet infrastructure and polluted water and soil. Since Baku remains an ethnically mixed city, however, and indeed home to many internally displaced persons from the Nagorno-Karabakh region, Slavs are not alone in enduring these environmental problems. Finally, in Azerbaijan's southern region near Lenkeran, Talechis, the dominant ethnic group, live in areas where the soil has eroded, the land is degraded, and coastlines are submerged. But, as in the northeast, these environmental conditions extend well beyond the region around Lenkeran.

In each of these three cases, the overlap between concentrated ethnic groups and environmental decline appears to be coincidental rather than intentional: environmental misfortune is experienced by a range of ethnic groups rather than befalling a particular one. What these patterns do not reveal is how people in these or other areas of Azerbaijan perceive their environmental conditions, if they associate their surrounding environment with health problems, or if environmental quality is even a top concern for them in their daily lives. In short, this mapping exercise raises further questions about how people integrate their perceived experience of environmental quality into their overall sense of well-being.

Human Security

This chapter takes the view that environmental injustice is part of a more complex set of living conditions that may be described as "human security." In contrast to traditional or neorealist approaches to security, which focus squarely on the state as the agent to maintain territorial integrity and provide security to citizens, and which do not generally consider the conditions of people's daily lives (Page and Redclift 2002), more recent approaches encourage a shift of focus to individual security in the broadest sense (Krause and Williams 1997). They question both the role of the state as the assumed provider of security and exactly what or who is being secured (Dalby 1997). There is a growing recognition of human security, with a focus on people rather than on states, in the field of security studies (Suhrke 1999; Hammerstad 2000; NSCP 2005).

Focusing on individual safety and freedom, rather than assuming these are assured to citizens by the state, raises two fundamental questions: "Security from what threats?" and "Security by what means?" (Bajpai 2000). Although national and human security should be mutually reinforcing, that over 90 percent of armed conflicts are intrastate suggests this is largely not the case and indicates a growing need to reassess how scholars and policy makers evaluate security (HSC 2005). State security, as it is traditionally conceptualized, cannot be presumed to provide citizens with day-to-day well-being. Not all states have the capacity to secure the well-being of their populations; in some cases, the state apparatus itself contributes to the *in*security of its citizens (Bellamy and McDonald 2002). As one scholar has noted, "Traditional security thinking has failed to deliver meaningful security to a significant proportion of the people of the world. This is an empirical reality. For most people, the greatest threats to security come from disease, hunger, environmental contamination, crime and unorganized violence. For many, a still greater threat may come from their own state itself, rather than from an 'external' adversary" (Newman 2004, 186). According to the United Nations Commission on Human Security, human *in*security may become a threat to state stability in instances where, for example, transnational terrorism, environmental pollution, massive population movement, and infectious diseases such as HIV/AIDS overwhelm populations and governments alike (UNCHS 2003, 5). Environmental instability and natural resource–related disputes have also emerged as concerns under the label "environmental security" and may have a direct impact on particular populations or regions (see, for example, Ascher and Mirovitskaya 2000).

On the other hand, when a state can protect and empower its citizens, human security and state security are indeed mutually reinforcing. It would seem reasonable, and some would argue ethical (Newman 2001), to expect that a state with the capacity to maintain or improve the basic elements of human security would do so, if only in the interest of ensuring its own legitimacy and stability. Like power (Allen 2004), security is not necessarily "locked" at one level or another. Instead of being isolated, dimensions of security are interrelated. Clearly, the state shapes the individual well-being of its citizens. State governments remain gatekeepers to economic development and the enhancement of human well-being through their pursuit of international aid and trade, their crafting of national economic policy, and their maintenance and improvement of education and health care systems as well as physical infrastructure. To understand human security, then, we need to understand how state-level pursuits and the day-to-day living conditions of citizens are interrelated.

This chapter's assessment of human security is guided by two basic measures that directly address day-to-day living conditions and carry implications for the longer time horizon—how well people are able meet their basic needs and how well the means to security are distributed (UNDP 1994; Newman 2001). Basic needs include a range of individual security concerns from food, health, economic, and personal security to community, political, and environmental security. Clearly, each of these aspects of individual security is greatly influenced by how well the means to achieve it—wealth, access to jobs, health care, education, infrastructure, and so on—are distributed. How well the benefits of natural resource use are distributed and to what extent environmental conditions hinder people's basic needs are important points where human security and environmental justice meet.

Impacts of Oil Wealth on Azerbaijan's Economy

Azerbaijan's oil wealth has been shown to have a defining influence on its development as a post-Soviet state (Karl 2000; Heradstveit 2001; Auty 2006); understanding the critical role of that wealth is key to understanding Azerbaijan's economy.

Table 4.1 demonstrates how significantly oil exports have increased since the late 1990s as Azerbaijan has developed its connections with the international oil industry. The volume of oil exported nearly trebled between 1998 and 2005, becoming the country's dominant export commodity. If the wealth generated by those oil exports had been evenly distributed among the Azerbaijani populace, each person's wealth, calculated as GDP per capita, would have also increased nearly threefold. However, that was emphatically not the case.

As in other resource-dependent, economically developing states, Azerbaijan's oil wealth is not trickling down to benefit most of the population. Oil drilling and refining have not stimulated parallel growth in other economic sectors where new jobs might be created. The status of Azerbaijan's agricultural sector has particularly significant implications

Table 4.1
Azerbaijan's oil exports and GDP per capita, 1998–2006

Year	Oil exports		GDP per capita (current international dollars)
	Percentage of total exports	Value (millions of dollars)	
1998	66	450	2,036
1999	78	801	2,198
2000	84	1,519	2,475
2001	90	1,841	2,764
2002	89	2,046	3,088
2003	86	2,250	3,477
2004	83	3,097	3,898
2005	90	4,989	5,027
2006	93	7,931	6,888

Sources: United Nations online database, http://data.un.org (accessed 8 September 2008), United States Energy Information Administration: Country Analysis Brief: Azerbaijan, http://www.eia.doe.gov/emeu/cabs/Azerbaijan/Full.html/ (accessed 4 March 2008).

for human security since it employs approximately 40 percent of the total workforce and potentially meets a basic need of the entire population. Yet agricultural productivity is declining: more people are producing less—thus generating and presumably dividing less wealth—per person (SSCRA 2004). The agricultural sector was traditionally strong in the Azerbaijani Soviet Socialist Republic; agricultural productivity in Azerbaijan actually increased between 1991, when the Soviet Union collapsed and Azerbaijan became independent, and 1999, when the country's oil industry began to take off (NIS/TACIS Services 2001, 28). More recently, though, low labor productivity in the agricultural sector has reflected a lack of efficiency, of effective reform, and of both domestic and foreign direct investment (EIU 2004, 27). Although agricultural productivity is again growing, at least slightly, the sector is still unable to meet domestic demand, generating a need for increased food imports. An influx of cheaper, imported food is likely to decrease further the competitiveness of the agricultural sector. Normally, as a country develops economically, agriculture expands through an increased use of machinery and technology, resulting in decreased human labor per unit of output. Agriculture in Azerbaijan, however, employs increasing numbers of people, suggesting that the slight increase in agricultural productivity is not a result of economic development or improved efficiency.

These trends in Azerbaijan's agricultural sector raise the question of wages, poverty, and the distribution of wealth throughout the country's economy. Overall, average monthly wages have been increasing. Table 4.2 shows trends over time in wages and infant mortality rate, a proxy for many indicators of human well-being and day-to-day quality of life (Harff 2003). Infant mortality rates have declined significantly since Azerbaijan became independent, and we would expect to see a significant corresponding rise in the country's standards of living. However, wages tell a more complex story. Real wages plummeted in the years following independence as Azerbaijan put its economic house in order; more recently, these wages have rebounded to previous levels and beyond. The increase in wages has not matched the increase in national wealth from oil exports, however. What is more, the growth in wages in recent years is unevenly distributed in favor of employees in the oil industry and related sectors—at the expense of the agricultural and social sectors (EIU 2004). Worse still, in 2002, despite the general upward trend in wages, nearly 47 percent of the total population fell well below an absolute poverty line (IMF 2004, 16–19). The Gini coefficient, a measure of income distribution, where 0 represents perfect equality and 1 represents perfect inequality, confirms these findings. The inequality in earnings distribution in Azerbaijan has been steadily increasing, from 0.275 in 1989 to 0.508 in 2002 (TransMONEE 2004).

Thus, instead of providing the means to improve economic conditions for the country's people, Azerbaijan's oil wealth is contributing to a greater economic imbalance among them. In turn, economic imbalance is likely to contribute to human insecurity within the country. Despite recent improvement in real wages and infant mortality rates, available figures for the Gini coefficient suggest that the gap between the haves and have-nots of Azerbaijan is widening.

From the perspective of environmental justice, the extent to which socioeconomic status correlates with concern about the environment is of interest here. As noted earlier in the discussion of the Kuznets curve, environmental degradation, in and of itself, may not generate public concern. Instead, only when people have sufficient resources, time, and security in other areas might they turn their attention to environmental issues. In an increasingly imbalanced economy such as Azerbaijan's,

Table 4.2
Wages and Infant Mortality Rates in Azerbaijan, 1989–2004

	1989	1990	1991	1992	1993	1994	1995	1996	1997	1998	1999	2000	2001	2002	2003	2004
Real wages[a]	100	101.1	80.0	95.0	62.4	24.8	19.8	23.6	36.2	43.3	51.9	61.2	70.8	83.5	100.3	120.6
Infant mortality	26.2	23.0	25.3	25.5	28.2	25.2	23.3	19.9	19.6	16.6	16.5	12.8	12.5	12.8	12.8	9.8

Sources: TransMONEE 2006 Database, UNICEF IRC, Florence, http://www.unicef-icdc.org/research/ (accessed March 4, 2008)
[a] Base year = 100.

we might expect correspondingly uneven perceptions of environmental conditions.

Public Concerns about Day-to-Day Life

To what extent are Azerbaijani citizens actually concerned about environmental issues? How do these concerns rank next to other concerns about day-to-day life in different places in Azerbaijan? Table 4.3 summarizes data collected in a survey of 1,200 Azerbaijani citizens.[2] Here, it is clear that environmental issues, in the form of pollution and access to natural resource–based utilities such as gas and water supply, rank among the top concerns of citizens across all of Azerbaijan. In every type of settlement, including large cities, small cities, villages, and even centers for internally displaced persons (IDPs), some form of environmental issue is a prioritized concern. In large and small cities, environmental pollution is a high-priority concern. In villages and IDP centers, respondents are less concerned with environmental pollution and instead rank gas or water supply as a high-priority environmental concern. These environmental concerns rank close to other high-priority items such as the Nagorno-Karabakh conflict, material well-being, and unemployment. In general, the data suggest that the public is aware of and concerned about the environment in different ways in different places.

Table 4.4 compares survey data about the respondents' economic situation and their concern about the environment. Respondents were asked about consumption rather than directly about income. Not only might people be reluctant to share income information with an outsider, they might be receiving income from different sources (e.g., salary, wages, pension, remittances), which could complicate their responses. Therefore, the survey instrument offered a range of responses tied to consumption as a proxy for economic status. The majority of respondents fall into a middle and lower economic status (29 percent and 31 percent of respondents, respectively), with very few people reporting an ability to purchase everything they need. About 20 percent of the respondents live in the most constrained economic conditions. Almost half of all respondents express extreme concern about the environment. That concern is relatively evenly distributed across socioeconomic levels and increases only slightly with economic status, showing that, among the sample population, environmental concern is not significantly correlated with a particular economic status. That finding differs from the work of Susmita

Table 4.3
Major concerns by priority and settlement type in Azerbaijan, 2006

	Large cities	Small cities	Villages	Internally displaced persons
High priority	Material well-being Unemployment Environmental pollution	Material well-being Unemployment Environmental pollution	Material well-being Unemployment Gas supply	Material well-being Water supply Gas supply
Medium priority	Access to health care Quality of education Crime Democracy and civil rights	Access to health care Gas supply Electricity supply Democracy and civil rights	Access to health care Quality of education Electricity supply Environmental pollution	Access to health care Quality of education Unemployment International relations
Low priority	Electricity supply Gas supply Water supply International relations	Quality of education Crime Water supply International relations	Democracy and civil rights Crime Water supply International relations	Democracy and civil rights Crime Electricity supply Environmental pollution

Source: O'Lear and Gray 2006.

Table 4.4
Economic status versus level of concern about the environment in Azerbaijan ($N = 1,200$)

Economic situation in your family?	How concerned are you about the environment?				
	Not at all	A little	Extremely	Do not know	Total
Not enough money for food	59	72	112	1	244
Difficulties with buying clothes	82	121	168	6	377
Expensive durable goods a problem	49	121	175	1	346
Automobile or trip abroad not possible for us	15	59	110	1	185
We can afford everything	2	5	6	0	13
Do not know or refuse to answer	10	4	21	0	35
Total	217	382	592	9	1,200

Dasgupta and David Wheeler (1997), mentioned earlier in this chapter, which showed that environmental concern in China was significantly greater among wealthier citizens. By contrast, economic status does not seem to influence the level of people's environmental concern in Azerbaijan.

As noted throughout this volume, a central characteristic of environmental injustice is the disproportionate rate of negative, environmentally related health impacts experienced by people of a lower economic status. Survey data may be used to assess whether such a distribution is evident among respondents in Azerbaijan. Table 4.5 compares economic status to negative, environmentally related health impacts. Most of the perception of negative health impacts clusters in the middle of the scale with about the same number of people claiming much, little, or no negative health impacts resulting from environmental pollution. Only 12 percent of the respondents perceive very significant negative health impacts in their family due to environmental pollution. The trend—although slight—appears to be just the opposite of what we might expect, with respondents in more difficult economic situations reporting *less* negative impact on their family's health than respondents with greater economic capacity. That is, instead of less well off people perceiving disproportionately greater harm from the environment, these data demonstrate that people who are *better* off actually perceive more significant, pollution-related health effects.

The survey data discussed so far in this chapter suggest that there is significant concern about the environment among people in Azerbaijan, including concern about pollution in larger settlements and concern about access to natural resource–based utilities in smaller settlements. Overall, concern with the environment is fairly evenly distributed across all economic levels. And, again counter to what we might expect when examining environmental injustice, the perception of negative health effects resulting from environmental pollution appears to be slightly stronger among people who are better off.

People who are more aware of the negative health impacts of pollution should logically also be more concerned about the environment. And, indeed, as the survey data displayed in table 4.6 show, public concern about the environment in Azerbaijan is directly related to the strength of perception of negative health impacts. Respondents who are more aware of environmentally related health problems within their family are more likely to be more concerned about the environment, and as the

Table 4.5
Economic status versus health impacts from pollution in Azerbaijan ($N = 1,200$)

Economic situation in your family?	To what extent does your family's health suffer because of environmental pollution?					Total
	Very much	Much	A little	Not at all	Do not know	
Not enough money for food	32	52	79	76	5	244
Difficulties with buying clothes	37	77	106	139	18	377
Expensive durable goods a problem	40	100	110	83	13	346
Automobile or trip abroad not possible for us	30	65	46	37	7	185
We can afford everything	0	6	5	2	0	13
Do not know or refuse to answer	6	6	8	10	5	35
Total	145	306	354	347	48	1,200

Table 4.6
Health impacts from pollution versus level of concern about the environment in Azerbaijan ($N = 1,200$)

To what extent does your family's health suffer because of environmental pollution?	How concerned are you about the environment?				
	Not at all	A little	Extremely	Do not know	Total
Very much	5	30	108	2	145
Much	6	71	228	1	306
A little	42	142	166	4	354
Not at all	158	112	76	1	347
Do not know	6	27	14	1	48
Total	217	382	592	9	1,200

awareness of these health problems decreases, so does environmental concern.

Conclusion

A premise of this volume is that the Soviet Union generated widespread environmental degradation without discrimination by wealth or ethnicity. The data presented here suggest that this premise does indeed hold true in Azerbaijan. Environmental concern appears to be widespread in Azerbaijan, albeit in different forms: public concern about the environment and awareness of environmental health effects are distributed evenly among most economic levels of the population. In the nearly twenty years since independence, Azerbaijan has undergone significant political and economic change, not the least of which is the rapid expansion of the oil industry and the wealth it is generating for the country. Yet most people in Azerbaijan have yet to see the benefits of that wealth in terms of widespread improvements in their physical living environments and personal economic prospects.

This chapter has drawn on the concept of human security to link environmental concerns to other aspects of day-to-day living conditions. It has presented evidence that the oil wealth in Azerbaijan—representing a great potential for the improvement of the lives and well-being of the Azerbaijani populace—is not yet being sufficiently channeled in such a way that most citizens there will benefit. Uneven growth in non-oil economic sectors and a growing income gap pose continuing challenges in the day-today lives of most Azerbaijanis. Those surveyed are predominantly concerned with securing material well-being and employment, namely, their ability to meet basic needs. Against this background of day-to-day human insecurity, concerns about environmental pollution and the reliability of natural resource–based utilities persist as related strands of environmental injustice.

The perception of environmental problems alone does not generate widespread or unified public participation focused on environmental justice. Indeed, table 4.3 demonstrates that Azerbaijanis share other, competing concerns, which may well take precedence over environmental issues. Yet it is not clear whether a culture of sustained public participation or dissent on any issue is developing in Azerbaijan. It may be that a significant change in economic conditions is necessary before people are motivated to engage in public activity such as dissent. For example, as Dasgupta and Wheeler's work (1997) suggests, citizens may be more likely to act on environmental concerns once their economic worries are significantly diminished. On the other hand, a worsening of economic conditions may lead people to feel that they have little to lose by engaging in public dissent or other forms of potentially risky involvement. The general public is well aware of the country's oil wealth; indeed, national slogans and imagery that promote the expanding oil industry may be building up public expectations of imminent benefits and improvements. But, though it is perhaps too early to judge the effectiveness of the State Oil Fund of Azerbaijan and other efforts to distribute benefits of oil wealth, thus far, economic conditions throughout the country show all signs of declining.

Related elements of human security (e.g., material well-being, aspects of distribution, environmental concern) may yet merge to influence public opinion and activity in Azerbaijan. In early January 2007, the Tariff Council announced immediate price increases for gasoline, natural gas, water, and electricity for home and institutional consumers alike. Speculation that these price hikes would lead to increased prices for other consumer goods such as bread and newspapers was quickly confirmed. The main political opposition, the Azerbaijan Popular Front Party, denounced the price hikes as destructive to the Azerbaijani people, called on the government to resign, and threatened to stage mass protests. In late January 2007, approximately 1,000 opposition supporters gathered in Baku to protest the recent price hikes. Baku municipal authorities had approved the protest, but two opposition websites were reportedly blocked, most likely to suppress public opposition. It remains to be seen what may actually result from these price increases in the long run. The immediate flurry of activity in their wake suggests, however, that in Azerbaijan, as in many other former Soviet republics, environmental justice is likely to have powerful economic components that merit close examination.

Acknowledgments

This chapter is based on work supported by National Science Foundation Grant no. 0514229 under the title "Geography of Environmental Conflict: The Case of Azerbaijan," which investigated why we do not see such conflict in Azerbaijan despite the presence of key indicators such as resource abundance, distributive scarcity, and ethnic diversity. Any opinions, findings, and conclusions or recommendations expressed in this chapter are those of the author and do not necessarily reflect the views of the National Science Foundation.

Notes

1. The environmental Kuznets curve is an inverted U-shaped relationship between environmental impact and wealth. The theory behind it posits that, as affluence of a country increases, usually through the process of industrialization, societal impact on the environment grows worse. In later stages of economic development, however, people become more concerned about the environment and have a greater financial capacity to lessen society's impact on the environment. In short, the environmental Kuznets curve suggests that environmental concern increases with wealth and not necessarily with environmental conditions. Studies of the environmental Kuznets curve differ in their conclusions, however. For a brief discussion of these studies, see O'Lear and Gray 2006.

2. Data were collected from a large-scale survey ($N = 1,200$) with key questions focused on people's day-to-day concerns, their perceptions and expectations of the international oil industry's operations in Azerbaijan, their environmental concerns (if any), and their degree and focus of political activity.

References

Agyeman, Julian, Robert D. Bullard, and Bob Evans. 2003. "Introduction: Joined-up Thinking: Bringing Together Sustainability, Environmental Justice and Equity." In Agyeman, Bullard, and Evans, eds., *Just Sustainabilities: Development in an Unequal World*, 1–18. Cambridge, MA: MIT Press.

Allen, John. 2004. "The Whereabouts of Power: Politics, Government, and Space." *Geografiska Annaler*, B, 86 (1): 19–32.

Ascher, William, and Natalia Mirovitskaya, eds. 2000. *The Caspian Sea: A Quest for Environmental Security*. Boston: Kluwer Academic.

Auty, Richard M. 2006. "Optimistic and Pessimistic Energy Rent Deployment Scenarios for Azerbaijan and Kazakhstan." In Richard M. Auty and Indra de Soysa, eds., *Energy, Wealth and Governance in the Caucasus and Central Asia: Lessons Not Learned*, 57–76. New York: Routledge.

Bajpai, Kanti. 2000. "Human Security: Concept and Measurement." Kroc Institute Occasional Paper 19:OP:1. http://kroc.nd.edu/ocpapers/op_19_1.PDF/ (accessed March 4, 2008).

Bellamy, Alex J., and Matt McDonald. 2002. "'The Utility of Human Security': Which Humans? What Security? A Reply to Thomas and Tow." *Security Dialogue* 33 (3): 373–377.

Costi, Alberto. 2003. "Environmental Protection, Economic Growth and Environmental Justice: Are They Compatible in Central and Eastern Europe?" In Julian Agyeman, Robert D. Bullard, and Bob Evans, eds., *Just Sustainabilities: Development in an Unequal World*, 289–319. Cambridge, MA: MIT Press.

Cutter, Susan L. 1995. "Race, Class and Environmental Justice." *Progress in Human Geography* 19 (1): 111–122.

Dalby, Simon. 1997. "Contesting an Essential Concept: Reading the Dilemmas in Contemporary Security Discourse." In Keith Krause and Michael Williams, eds., *Critical Security Studies: Concepts and Cases*, 3–31. Minneapolis: University of Minnesota Press: Pinter.

Dasgupta, Susmita, and David Wheeler. 1997. "Citizen Complaints as Environmental Indicators: Evidence from China." Working Paper no. 1704. Infrastructure and Environment Development Research Group, World Bank. http://ideas.repec.org/p/wbk/wbrwps/1704.html/ (accessed March 4, 2008).

DeBardeleben, Joan, and John Hannigan, eds. 1995. *Environmental Security and Quality After Communism: Eastern Europe and the Soviet Successor States.* Boulder, CO: Westview Press.

Economist Intelligence Unit (EIU). 2004. *Country Report: Azerbaijan.* November. London.

Eyerman, Ron, and Andrew Jamison. 1991. *Social Movements: A Cognitive Approach.* University Park: Pennsylvania State University Press.

Gulbrandsen, Lars, and Arild Moe. 2007. "BP in Azerbaijan: A Test Case of the Potential and Limits of the CSR Agenda?" *Third World Quarterly* 28 (4): 813–830.

Hammerstad, Anne. 2000. "Whose Security?: UNHCR, Refugee Protection and State Security after the Cold War." *Security Dialogue* 31 (4): 391–403.

Harff, Barbara. 2003. "No Lessons Learned from the Holocaust? Assessing Risks of Genocide and Political Mass Murder since 1955." *American Political Science Review* 97 (1): 57–73.

Heradstveit, Daniel. 2001 *Democracy and Oil: The Case of Azerbaijan.* Wiesbaden: Reichert.

Human Security Center (HSC). 2005. "The Human Security Project Report." University of British Columbia. http://www.humansecuritycentre.org/ (accessed March 4, 2008).

Karl, Terry Lynn. 2000. "Crude Calculations: OPEC Lessons for the Caspian Region." In Robert Ebel, and Rajan Menon, eds., *Energy and Conflict in Central Asia and the Caucasus*, 29–54. Lanham, MD: Rowman and Littlefield.

Krause, Keith, and Michael C. Williams. 1997. "From Strategy to Security: Foundations of Critical Security Studies." In Keith Krause and Michael Williams, eds., *Critical Security Studies: Concepts and Cases*, 33–59. Minneapolis: University of Minnesota Press.

Micklin, Philip P. 1992. "The Aral Crisis: Introduction to the Special Issue." *Post-Soviet Geography* 33 (May): 269–282.

Newman, Edward. 2001. "Human Security and Constructivism." *International Studies Perspectives* 2: 239–251.

Newman, Edward. 2004. "The 'New Wars' Debate: A Historical Perspective Is Needed." *Security Dialogue* 35 (2): 173–189.

New Security Challenges Program (NSCP). 2005. Core Script. University of Birmingham. http://www.newsecurity.bham.ac.uk/ (accessed March 4, 2008).

NIS/TACIS Services. 2001. *Economic Trends Quarterly Issue: Azerbaijan, July–September 2000*. Brussels: European Commission.

O'Lear, Shannon. 1996. "Using Electronic Mail (E-Mail) Surveys for Geographic Research: Lessons from a Survey of Russian Environmentalists." *Professional Geographer* 48 (2): 213–222.

O'Lear, Shannon. 1999. "Networks of Engagement: Electronic Communication and Grassroots Environmental Activism in Kaliningrad." *Geografiska Annaler* 81: 165–178.

O'Lear, Shannon. 2007. "Azerbaijan's Resource Wealth: Political Legitimacy and Public Opinion." *Geographical Journal* 173 (3): 207–223.

O'Lear, Shannon, and Angela Gray. 2006. "Asking the Right Questions: Environmental Conflict in the Case of Azerbaijan." *Area* 38 (4): 390–401.

Page, Edward A., and Michael Redclift, eds. 2002. *Human Security and the Environment: International Comparisons*. Cheltenham, UK: Edward Elgar.

Pellow, David Naguib, and Robert J. Brulle. 2005. "Power, Justice, and the Environment: Toward Critical Environmental Justice Studies." In Pellow and Brulle, eds., *Power, Justice, and the Environment: A Critical Appriasal of the Environmental Justice Movement*, 1–22. Cambridge, MA: MIT Press.

Peterson, D. J. 1993. *Troubled Lands: The Legacy of Soviet Environmental Destruction*. Boulder, CO: Westview Press.

Rasizade, Alec. 2003. "Azerbaijan in Transition to the 'New Age of Democracy.'" *Communist and Post-Communist Studies* 36: 345–372.

Rhodes, Edwardo Lao. 2003. *Environmental Justice in America: A New Paradigm*. Bloomington: Indiana University Press.

Ringquist, Evan J. 2006. "Environmental Justice: Normative Concerns, Empirical Evidence, and Government Action." In Norman J. Vig and Michael E. Kraft, eds., *Environmental Policy: New Directions for the Twenty-First Century*, 6th ed., 239–263. Washington, DC: CQ Press.

Saiko, Tatyana. 2001. *Environmental Crises: Geographical Case Studies in Post-Soviet Eurasia*. Harlow, UK: Prentice Hall.

State Statistical Committee of the Republic of Azerbaijan (SSCRA). 2004. *The Statistical Yearbook of Azerbaijan 2004*. Baku. http://www.azstat.org/publications/yearbook/SYA2004/indexen.php/ (accessed March 4, 2008).

Suhrke, Astri. 1999. "Human Security and the Interests of States." *Security Dialogue* 30 (3): 265–276.

TransMONEE. 2006. Database, UNICEF IRC, Florence. http://www.unicef-icdc .org/research/ (accessed March 4, 2008).

United Nations Development Program (UNDP). 1994. "Human Development Report 1994: New Dimensions of Human Security." Human Development Report Office, New York. http://hdr.undp.org/reports/global/1994/en/default.cfm/ (accessed March 4, 2008).

United Nations. Database, http://data.un.org (accessed 8 September 2008).

United States Energy Information Administration (UNEIA). "Country Analysis Brief: Azerbaijan." http://www.eia.doe.gov/emeu/cabs/Azerbaijan/Full.html/ (accessed 4 March 2008).

Yanitsky, Oleg. 1993. *Russian Environmentalism: Leading Figures, Facts, Opinions*. Moscow: Mezhdunarodnye Otnosheniia Publishing House.

5

Civil Society and the Debate over Pipelines in Tunka National Park, Russia

Katherine Metzo

Billionaire oil oligarch Mikhail Khodorkovsky was convicted of fraud and tax evasion in 2005 and sentenced to nine years in jail.[1] His company, Yukos Oil, had proposed the Angarsk-Daqing pipeline, which would transport oil from the Kovykta Field in Irkutsk Oblast to the refinery city of Daqing in northeastern China. By late 2003, those plans were put on hold when the government returned the company's environmental impact assessment for revisions. On the heels of this disappointment, Khodorkovsky was arrested for alleged tax evasion, falsifying documents, and theft. His arrest was unconnected to the pipeline, with some analysts arguing that Putin had targeted Khodorkovsky for supporting Putin's political opponents (Pazderka 2005; Goldman 2004).

There is another, less-known account of why Khodorkovsky was arrested, that supernatural forces had been mobilized against him to preserve the Tunka Valley, a protected area located along his proposed pipeline route. Natalia Zhukovskaia, a prominent Moscow ethnographer who has spent forty years researching the Tunka Valley, suggested in an interview with the newspaper *Nomer-odin* (Number One) that Khodorkovsky's arrest and jail sentence had been influenced by a group of shamans from the Tunka Valley. In late 2002, at the height of local protests over the proposed construction of a pipeline through their region, the shamans performed a ritual. "No federal or regional bureaucrats could scare [the shamans or Buddhist lamas]," Zhukovskaia explained, "since they have their own authorities: spirits, gods, and sacred places of Buriatiia." While local residents wrote letters to President Putin and Prime Minister Kasyanov, "[the] shamans, who don't believe in the strength of the written word, conducted a series of prayers near the most sacred places of the valley. What followed is well known [*Dal'neishee vsem nam izvestno*]" (in Samojlova 2005, 8).

That a shamanic ritual had been responsible for Khodorkovsky's arrest may seem preposterous, especially given the openly hostile relationship between Vladimir Putin and the Russian oligarchs. Nevertheless, religious ritual was one strategy of protest against the oil pipeline. Although Zhukovskaia herself has acknowledged the more pragmatic explanations for the pipeline's failure, her intent was to draw attention to the dire socioeconomic consequences of pipeline construction for the livelihoods of indigenous peoples, an issue ignored during the political intrigue surrounding the case (see Zhukovskaia 2004).

This chapter examines two proposed pipelines from the 2000–2003 period that were slated to run south of Lake Baikal through the Tunka Valley. The first, proposed in 2000, was a natural gas pipeline to China longer than the Alaskan Oil Pipeline. Although this project raised little overt protest, and indeed still has some supporters in the region, the introduction of the Yukos project a year later increased local protest against pipelines in general. As of July 2007, neither pipeline had been approved for construction through the Tunka Valley, and Yukos's assets had been auctioned off, but the natural gas pipeline proposal has again gained momentum. A key reason for local opposition is that both pipeline routes would have cut across protected zones adjacent to Lake Baikal including "specially protected zones" such as the Tunka Valley, home to a national park.

One of the questions this chapter seeks to answer is why one pipeline proposal was met with local resignation and even apathy, whereas the second raised widespread concern and prompted social action instrumental in halting the planned incursion into this protected landscape. I suggest that part of the difference lay in the different perceptions of power, responsibility, and accountability that locals had about the two proposed pipelines. In the case of the natural gas pipeline, which was initially understood as a state-led initiative, people had one of two responses.[2] Some were less concerned about pipeline construction because they perceived the state as ineffectual and lacking sufficient finances to proceed with the project. Others viewed the government as a legitimate authority to approve a pipeline, but they recognized that the state also had a legal duty and moral responsibility to safeguard the integrity of both the park and Lake Baikal as protected sites.

With the introduction of the Yukos proposal, the threat of a pipeline on the territory of the Tunka Valley seemed imminent. Indeed, a pipeline through this ecologically fragile and seismically active territory would directly threaten both the integrity of the ecosystem and the livelihoods of the local population. It was the perceived immediacy and magnitude of this threat to their livelihoods that finally propelled local residents into action. The Yukos pipeline also mobilized an international community of nongovernmental environmental organizations. In fact, I first learned of the Yukos proposal in Spring 2002, when my college roommate forwarded an environmental action alert from Global Response. The alert focused first on protecting Russia's boreal forest ecosystems as a whole—noting that Tunka represents a "pristine taiga forest ecosystem" (Global Response 2002)—and the unique Baikal ecosystem in particular. It then went on to address the impact of the pipeline on local indigenous residents. "Yukos plans to build the pipeline within 200 meters (660 feet) of specific sites that are sacred to the Buriats, who suffered persecution under Stalin for their traditional practices of Buddhism and shamanism" (Global Response 2002). However, in raising the issue of environmental racism, the targeting of a particular minority ethnic population, it failed to mention the impact on the Russians and other Slavs who make up almost 40 percent of the population of Tunka.

Global Response focuses on letter writing as a means of "democratizing" and internationalizing local environmental justice movements. As nonspecialists in the local debate, its staff members rely on local NGOs for framing the issues (Palmer 2005), but it is clear from the text of the Tunka action alert that the local NGO was simply not familiar with the ethnic composition of the region. And however important the threats to sacred sites are to locals, a more pressing concern is the impact the pipeline would have on their livelihoods as sheep and cattle herders, hunters, fishers, and foragers (see also Peña 1992, 2005a).

By placing the marginalized rural population of Tunka at the center of the debate on pipelines, activists have clearly situated the two pipeline proposals within the model of environmental justice (Agyeman 2005; Harvey 2000; Johnston 1997). Population density is low in the national park, as in the case of Alaskan natives. Although most households combine their traditional livelihoods with some kind of wage labor, nomadic animal husbandry, hunting, fishing, and the collection of nontimber forest products (see Metzo 2001) all require both a great deal of land and that the ecological integrity of this land be sustained. Thus the impact of any invasive project in the region upon the local population is of utmost importance in any environmental assessment (see Johnston 2001).

A second, and equally important, question this chapter seeks to answer is how people *responded* to this threat. In answering, the chapter will highlight the complex network of relationships between locals and a host of actors in locations around Russia and the world. As a social movement, the local response to the pipeline was neither cohesive nor clearly articulated. Resource mobilization theory suggests that this is not atypical: collective action within social movements is most often dominated by interest-group politics, rather than by those most disaffected (Edelman 2001). The environmental justice literature likewise suggests a movement whose goals appeal to a broad base (Agyeman 2005; Pellow and Brulle 2005), but which struggles with achieving "just" or "sustainable" outcomes (Toffolon-Weiss and Roberts 2005; Fortwangler 2003). Although there were clearly interest groups and outside individuals who performed significant work in the Tunka pipeline protests, the indigenous people were equally involved in activities with direct and indirect impacts. Some, as members of an elite subgroup of indigenous leaders from throughout Siberia, met with Alaskan natives and environmental activists. Others wrote letters and engaged in protests at public hearings. Still others—the shamans—engaged in ritual protest as a culturally specific form of collective action.

Though there is a consensus that the Soviet Union lacked a civil society (Evans 2006), anthropological inquiry paints a more complicated picture (see Hann 1995; Phillips 2005). Regarding the role of civil society in the collapse of Communism, Chris Hann remarks, "Radical opposition to socialism was restricted to small, politically conscious fragments of populations and it should also be remembered that many of those who struggled to change Communist systems did so from the space they managed to find within the state" (Hann 1996, 9; see also Buck-Morss 2000, esp. chap. 6). By and large, the literature on both civil society and social movements focuses too narrowly on organizations and planned activities. Environmental justice literature has been more effective at recognizing the constellation of related activities that make up "social movements."[3] I argue that framing the debate in terms of threats to traditional livelihoods comes from within Tunka itself, where the discourses of conservation and livelihoods have been in tension for many years.

As the title of this chapter suggests, the debate over the Tunka pipeline projects does indeed represent the mobilization of civil society, keeping in mind that civil society, social movements, and resistance are not unified, homogeneous actions (Gupta and Ferguson 1997; Hann 1996). In the case of Tunka, civil society operates, without clearly framing itself, as a loosely connected environmental justice movement.

Tunka National Park

The Tunka Valley is situated some 110 kilometers (68 miles) southwest of Lake Baikal and is home to 26,000 residents living in over 30 small villages along the Irkut River corridor. As a representative of the Altai-Sayan eco-region and home to over two hundred mineral springs, the valley is also a natural conservation area, Tunka National Park, which protects dozens of natural, cultural, and historical monuments. At 1.2 million hectares (3 million acres, or 4,630 square miles), it is the largest national park in Russia. To me, Tunka, as the locals refer to it, is also a sleepy rural area of Siberia, near Lake Baikal, where people live their lives in much the same way they did before the park was formed in 1991.[4] Unlike their counterparts in many national parks throughout the world, local residents in Tunka were not displaced by the park and participated in early discussions about protection and management. They have jobs, tend gardens and livestock, visit family and friends on weekends, drink healing mineral waters, and take trips to the forest to harvest berries and mushrooms at peak season. Typical of rural Russia, Tunka has been hit with high unemployment and extreme poverty. Families depend on activities like foraging as part of their survival. The park and related tourism industry are critical to providing salaried employment and supplemental seasonal income. Therefore, the conservation status of the park is a crucial consideration for locals.

The borders of the park, which are coterminous with those of Tunka administration district, have been under debate since its inception. Some have proposed carving two smaller parks, one in each mountain range, out of the greater part of Tunka and leaving the settlements along the valley floor open to a wider range of economic development possibilities. Although this alternative leaves much of the ecosystem intact, it focuses attention on a corridor that is already partially developed. If a pipeline were to go through the valley, however, the boundaries would have to be altered so that the pipeline would transect both the Sayan and the Khamar Daban ranges. Any boundary changes would have to be modified through legislation. In 2003, pipeline proponents backed such a move, stating that the original boundaries of Tunka National Park were ambiguous, having been arbitrarily, albeit intentionally (see Metzo 2003), drawn to match those of the administrative region. Persistent legal petitioning by local activists has blocked attempts to clarify the park's boundaries.

The pipeline proposal and local protests were not anything I could have anticipated when I embarked on my research into household economic behavior within a protected area in 2000 and 2001. However, because I spent time at the national park offices conducting interviews, collecting data on resource management, and consulting on ecotourism development, macro-level concerns like the pipeline also came to light. I included information and conversations about the pipeline in my journal, but there was relatively little public discussion of the issues in 2000 and 2001, for reasons discussed below. Not being present to observe protest activities in 2002, I have used online resources and news accounts that mentioned the proposed pipelines to reconstruct events. The story that emerged became the foundation for discussions with consultants in Tunka when I returned for a new project in 2005. Access to the local news archive and informal discussions with the newspaper editor helped round out the picture.

A Brief Timeline of the Pipeline Controversy

The earliest discussion of a pipeline was the mention of the government's intentions to develop a natural gas resource in adjacent Irkutsk Oblast and run a pipeline to or through Tunka.[5] Valerij Ivanovich Tolmachev, state ecological inspector for Tunka, mentioned the pipeline in 2000 during a public report on energy pollution statistics in two villages. Municipal buildings in the two largest villages, Kyren, the administrative center, and Arshan, a resort town, heat with coal. Private residences use firewood from local forests. Arshan's microclimate contributes to its air pollution problem by trapping particulates rather than allowing them to disperse. Raising the question of this worsening problem, Tolmachev brought up natural gas, hoping to explore the pros and cons. No one considered a natural gas pipeline a viable alternative because of the environmental costs of construction through seismic areas and bog habitat. A single accident, some argued, could pollute many or all of the natural mineral springs and village wells, which are the only water source for locals and visitors.

My initial, flawed understanding of this pipeline was that it was a branch pipeline running into Tunka to provide a cleaner fuel source. From Tolmachev's perspective, the pipeline, which he agreed was not the best alternative, was a way to open up a dialogue about alternatives to coal. Fifteen people participated in this meeting and when I raised the question about a pipeline outside this small group of concerned citizens, most residents expressed little concern. Instead, they threw up their hands and dismissively stated that the government would do what it wanted. The group planned to meet in two weeks to discuss the energy issue further, but that meeting was preempted by other, more immediate concerns, such as a conflict over a local Buddhist temple. Later, in summer 2001, when Yukos conducted its feasibility study, the region faced a major natural disaster—snowstorms and mudslides in the mountains had stranded ecotourists and destroyed homes—so local discussion of the pipeline was put off once more.

Soon after my departure in August 2001, however, the oil pipeline proposed by Yukos evoked strong protests among Tunka residents. In October 2003, I located a document on a "direct democracy" Web site in which a representative of Baikal Environmental Wave (BEW), an Irkutsk-based NGO, notes that during public hearings for the oil pipeline in Tunka, "the public discussion of the oil pipeline project in Kyren is so far the only example of how these discussions should work and how the public *should* act." After twenty formal speeches, uncounted informal comments, and at least fifty questions, much of it in people's native Buriat language, it was obvious that "Yukos' representatives did not feel so sure of themselves anymore" (Belskaya n.d.; emphasis added).[6] Those in attendance who opposed the pipeline refused to end the public meeting until a "zero option" (rejection of the project—no pipeline) was included in the final document.

Environmental Justice, Civil Society and Temporary Success in Tunka

If civil society is defined as a response to despotism (see Hall 1995), then the people of Tunka indeed acted as they "should act" in public hearings. The despot in this case, however, was a capitalist oligarch rather than a tyrannical leader. And protests against the Yukos pipeline were generated by a number of independent actors coming together in a seemingly spontaneous way rather than by a unified social movement. Until this point, my focus has been on international, urban-based, environmental groups because they were the means by which I learned about the pipeline proposals. For Tunka residents, however, the debate about pipelines draws on familiar conflicts about autonomy (Peña 2005a).

Elements associated with civil society—NGOs and the media as well as religion and social networks—have informed and framed the environmental justice movement generally, and the environmental justice debate over pipelines in Tunka more specifically. Julian Agyeman (2005, 26) identifies three types of justice sought in environmental justice movements: procedural justice, which might be referred to as "genuine participation in planning"; substantive (or material) justice; and distributive justice, focused on equity in sharing ecological and economic benefits.

Two issues, in particular, mark the Tunka pipeline debate. First, forms of social consciousness from the Communist past have been mobilized in original ways. In some cases, institutions developed in the late Soviet era have helped to buffer people of the former Soviet republics against the harsh realities of the capitalist economy (see Ninetto 2005). Even more striking in the Tunka case, however, are the social networks created under Soviet Communism, through organizations such as the Komsomol and state universities, that have trained the people of Tunka in the organizational skills they need to succeed within this post-Soviet context and have provided them a base from which to mobilize others (Yurchak 2003; Phillips 2005; Buck-Morss 2000).

Second, conservation plays a paradoxical but central role in the Tunka discourse about people and place. Although parks and conservation are often posed as contradictory to the continuation of indigenous livelihoods (Igoe 2004), conservation is essential to the preservation of a "traditional way of life" in Tunka (cf. Brechin et al. 2003). In the late 1980s, many considered conservation a threat to traditional lifestyles, in particular because park status allowed the state to prohibit precisely those activities traditional peoples depended upon for their survival (Metzo 2009). That the traditional way of life discourse today depends upon the existence of protected lands calls to mind Foucault's discussion of discourse as a "tactical element" within social movements (see Foucault 1978, *The History of Sexuality* as applied to political movements by Gupta and Ferguson 1997, 18–19).

Organizations

There are several different types of environmental organizations involved in the debate over pipelines in Tunka (see table 5.1). Although interna-

Table 5.1
Environmental organizations involved in Tunka pipeline debate

International	Greenpeace
	Initiative for Social Action and Renewal in Eurasia (ISAR)[a]
	Taiga Rescue Network[a]
	Pacific Environment[a]
	Global Response
	Green Cross
	World Wildlife Fund (WWF)
Russian or regional	Baikal Environmental Wave[a]
	Buryat Regional Department on Lake Baikal[a]
Local	Zov Arshana (Call of Arshan)[a]
	Akhalar (Buddhist ecological organization)[a]
	Sayany

[a] Attended public hearings in Tunka or at other sites along proposed pipeline routes.

tional NGOs such as Global Response are the most numerous type, because their presence or absence was little noticed by local residents, I suggest that they are less important to the debate on pipelines than local NGOs.[7] A photograph accompanying one of the Internet reports on the Tunka Valley and the Yukos pipeline showed a group of national park staff workers. When, in 2005, I told them how impressed I was that they had sat through public meetings lasting over twelve hours, the three colleagues present all smiled.[8] One laughed, then paused before replying, "Well, we weren't in any hurry [*Nam nekuda speshit'*]." (The public hearings were held in mid-July, between planting and harvest seasons, when rural dwellers do indeed have more time on their hands.)

Laura Henry (2006) presents a neat typology of "government-affiliated," "professionalized," and "grassroots" environmental organizations in several regions of Russia. As testimony to its complexity, however, her categories do not seem to fit the Tunka situation. The mismatch, in part, reflects the dearth of local environmental organizations, on the one hand, and the extended reach of urban environmental organizations, on the other, a situation that is not unusual for the environmental justice movement more generally.

Government-affiliated organizations receive government financing and organizational support. The only such organization in this case is the Buriat Regional Department (BRD) on Lake Baikal, led by Sergei

Shapkhaev. The leaders of these organizations, according to Henry (2006, 221), tend to work within administrative posts, to "use government-friendly rhetoric, and [to] argue that environmental protection is a state function that society should support, not challenge."

An exception to this tendency, Shapkhaev is an academic with a long history of working with both the government and NGOs in the Lake Baikal World Natural Heritage Site. At his university, East Siberian State Technical University, he and colleagues have founded departments and programs dedicated to the scientific and social aspects of sustainable development. Like many indigenous academics, Shapkhaev combines his scientific research with a strong set of Buddhist spiritual values. In stark contrast to Henry's government-friendly leaders, Shapkhaev earned the honor of runner-up Environmentalist of the Year from Condé Nast in October 2005, in large part for his lawsuit *against* the government for proposing to change the boundaries of Tunka National Park to accommodate the Yukos pipeline. Yukos officials tried to negotiate with Tunka residents to gain support for a partnership that would include altering the boundaries of the park (see also Stammler and Wilson 2006). Shapkhaev and colleagues conducted public opinion surveys in Tunka, organized public hearings, and helped to collect signatures on a petition against changing park boundaries. The proposed boundary changes were a direct violation of the federal law that established the park in 1991, which gave BRD a rock-solid foundation for the lawsuit. At the same time, Shapkhaev and colleagues went to work in Irkutsk Oblast, where the route proposed for the northern oil pipeline cut across traditional Evenk grazing territories (Fondahl and Sirina 2006).

The Irkutsk-based regional organization Baikal Environmental Wave (BEW) exemplifies Henry's second category, "professionalized" organizations, whose leaders are typically academics. BEW, led by British national Jennie Sutton, an activist who works closely with regional academics, has collaborated with Shapkhaev's BRD to sponsor public hearings in the affected regions and to press Yukos and Transneft to make their environmental impact assessments available to the local population in advance (usually with limited success by Sutton and Shapkhaev's accounts). Baikal Environmental Wave has widened its scope from strictly nature conservation, such as the illegal hunting of *nerpa*, Baikal's native seal, to include sustainable livelihoods (Sutton 2003; cf. Brechin 2003). Because of her longtime residence in the region and her advocacy of Lake Baikal, Sutton is considered by many to be a local herself.

Grassroots organizations, Henry's third category, which she describes as local organizations with a small or nonexistent budget, tend to be led by educators. Accordingly, they also tend to be apolitical, focusing on local resource issues and educational priorities within the community. Given its active role in educational initiatives in collaboration with the Lyceum School in the village of Arshan, the environmental group Zov Arshana (The Call of Arshan) fits this category. However, even before the pipeline controversy, members of Zov Arshana were active in patrolling the forests around their town to document illegal timber harvesting. In 2001, Zov Arshana was awarded a small grant from Global Green to monitor the planning and development of the Yukos pipeline (http://www.greengrants.org/cgi-bin/grants.cgi/ [accessed February 16, 2004]). Grassroots organizations engage with the community publicly and formally, through hearings, and informally, through kinship networks and social networks of former classmates, students, or coworkers, as discussed below.

Media

Another key element framing the Tunka protests is the media, especially newspapers. Because small regional newspapers are so widely read by locals, they can be an effective means of communication in Russia. Anne White (2006) notes that, even though circulation for such newspapers is in decline in regions throughout Siberia, actual readership may not be. As one informant told her: "It's my own [*rodnaia*]: how could I not read it?" (White 2006, 291). In Tunka, municipal offices generally receive at least one copy of the local weekly newspaper *Sayany*, and often multiple copies if the office is divided into departments. Employees in these offices generally take turns reading through the paper, as do visitors to the park offices, while waiting to speak with whomever they came to see. Home-delivered copies also have multiple readers, both because households are often large and because friends and relatives will often visit one another in the evenings to watch a favorite television program. I estimate that, conservatively, from four to six people read each newspaper delivered in Tunka.

Thus newspapers are highly effective in informing a population, helping to create a sense of shared identity (Anderson 1983), and providing direct and immediate communication on key social problems (see White 2006; Edelman 2001). In the post-Soviet period, the question of bias creeps in, however: many newspapers depend on the money they receive to publish articles written by industry. And so it came as no surprise that Baikal Environmental Wave encountered resistance from major city newspapers in Irkutsk Oblast when it attempted to publish information about public hearings on the Yukos pipeline.

The commercial use of newspapers contrasts the news philosophy of Vladimir Tulaev, the editor of *Sayany* until 2004. Tulaev views the role of a regional newspaper to be a public forum for important local issues. The column "Pipelines through Tunka: Yes or No?" ran virtually every week of 2002, with a brief hiatus for holidays and preelection interviews. In all, there were sixty-two letters and articles submitted by a diverse audience of local residents: hunters, national park staff, schoolchildren, pensioners, religious leaders, and even an industry spokesperson (for whose statement the paper received no payment). Although Tulaev strove to balance pro and con in putting together the column each week, his first commitment was to allow every voice that wanted to reach the newspaper's audience to be heard. The vast majority of articles were opposed to the pipelines. Indeed, according to a 2003 survey conducted by BRD, two-thirds of the population of Tunka was opposed to the Yukos pipeline (Hengesbach and Shapkhaev 2003, 18). How did he achieve balance then? I asked. Did he have to turn away letters or articles? He responded that, although "maybe one or two" that were not really on topic or did not take a position were sent back to the authors, his strategy for maintaining a balance between pro and con if, say, there was only one submission in favor of the pipelines was to delay publication of an opposition submission for a week or two.

A thorough analysis of the letters and articles in the for-and-against pipeline column is beyond the scope of this chapter, but a preliminary analysis highlights several key arguments on each side. Those in favor tended to discuss the benefits of clean-burning natural gas, even after the focus of the debate shifted to the Yukos oil pipeline proposal, to point out the environmentally friendly nature of the technology that would be used for pipeline construction, and to note the increase in local employment such construction would bring.

Those opposed to the pipelines argued that the environmental records of Russian energy companies did not match their environmentally safe construction promises, that most jobs on the pipeline required specialized training and would go to outsiders, and that any benefits from the oil pipeline would be short-lived, based on its 25-year life expectancy, and therefore not sustainable. Hunters argued that pipeline construction would irrevocably impact animal habitats in traditional hunting areas, thus putting their livelihoods in jeopardy. Several authors invoked a Bruntland-type definition of sustainability—meeting the needs of today's generation without eliminating the possibility of future generations to meet their own needs—but focused on what could be called "autonomy in livelihood" (Peña 2005a, 2005b), rather than on environmental justice or sustainability as recently understood (see Agyeman 2005).

Religion
The shamanic ritual mentioned at the beginning of the chapter was meant to prevent any pipeline from being built on the protected territory. Budazhap Shiretorov, a shaman from Tunka, explains: "When we go to a sacred site, we bring offerings [*prepodnosim podnosheniia*], and we offer them to the master spirits of the mountains and rivers" (Samojlova 2005, 8). Even national park staff make proper offerings when visiting sacred sites in the course of their work. The consequences of not appealing to the master spirits of sacred places, of which there are hundreds in Tunka, are harm and even catastrophe, in this case not only to those building the pipeline, but to all the residents of the region (Samojlova 2005; Shaglanova 2007). Shamans, then, can be seen as voluntarily engaging in collective action, a ritual, to the benefit of others in society. In addition to their religious role, many shamans also work at part- or full-time jobs.

Buddhist monks played a similar protective role, as the following assessment of the Yukos public hearings in Kyren makes clear:

Here the clash of two worldviews was vividly illustrated. On the one side, villagers and Buddhist monks, speaking in their own language, expressing a lack of trust towards YUKOS's promises, fears that incoming workers would not respect the natural environment, fear for their national park and sacred places, traditional way of life, and the forest with all it gives the local people and on which they depend. On the other, a polished Muscovite expressing distress and shame at the fact that his fellow citizens should make a living out of hunting, fishing, picking berries, nuts and mushrooms, and a plot of land. (Sutton 2003)

Although the account of this "clash" is instructive, I want to draw attention to "speaking in their own language" during a public hearing. All the Buddhist monks in Tunka and virtually all residents, except perhaps the very elderly, speak Russian as a second language. For the Buddhist

monks, speaking the Buriat language instead of Russian, which would
have been intelligible to the "polished Muskovites," is an overtly political act.[9] In speaking Buriat, meeting participants emphasize cultural difference to highlight the livelihood they are speaking out to protect.

Social Networks

As they were in the Soviet Union, social networks have been an effective
means for accessing information, goods, and services in post-Soviet
Russia (Ledeneva 1998; Pesmen 2000; see also White 2006). The social
consciousness and organizational skills used by many activists are also
derived from Soviet practices (Yurchak 2003). Sarah Phillips (2005)
demonstrates that female leaders of nongovernmental organizations in
Ukraine often take up personally meaningful causes, especially issues
where they feel the government has failed to protect them.[10] Yet they de-
ploy the organizational strategies learned as members of Komsomol or
other Soviet youth organizations. Likewise, many of the social networks
and organizational strategies operating in the pipeline debate in Tunka
draw both on new relationships to international NGOs and on Soviet
education and organizational training.

In running the for-and-against series in *Sayany*, Vladimir Tulaev was
continuing a practice he had established already in the 1980s. Tulaev's
concern for ecological questions is long standing; he is also a member of
a regional writers' guild to which the two cofounders of Tunka National
Park, Vladimir Syrenov and Ardan Angarkhaev, also belong. Angarkhaev, along with Sergei Shapkhaev, director of the BRD, and other academics and activists, later worked on establishing the Lake Baikal World
Natural Heritage Site. Another activist for this cause was Lama Tenzin
Khetsun Samaev, a Buddhist monk and spiritual teacher for the three
Buddhist temples operating in the Tunka and Oka regions. Lama
Samaev had both a religious and secular education, and was the founder
of the Buddhist ecological organization Akhalar. Through their work on
creating a World Heritage Site, a range of individuals tied to pipeline
protests were linked to an international audience of environmental
organizations well before the introduction of any pipeline proposal.

Of numerous trans-local social network links, I will briefly mention
two. A notable, albeit indirect, connection to the pipeline debate is
American Dan Plumley, whose Totem People's Preservation Project
(http://www.totempeoples.org) operates in Oka, located directly northwest of Tunka and also affected by the pipeline. Plumley first came to

the region as part of the international team conducting a formal study of the Baikal region for World Heritage status, and he has maintained ties to Shapkhaev and the staff at Tunka National Park. He assisted the park staff with international exchanges in the United States and has supported Lama Samaev's efforts to obtain federal minority status for the Soyot reindeer herders with whom he works. This later effort also links him to ethnographer Natalia Zhukovskaia, long an ally of local leaders in Tunka and Oka, having worked in the Sayan Mountains for forty years. In addition to interviews like the one that opened this chapter, she has written internal documents on the potential social impact of the pipeline on the local people.

There are many other nodes and connections that could be articulated, but this brief outline amply demonstrates that the relationships between actors were already well established before they participated in the pipeline protests. Several relationships date back to the Soviet era and to common interests in literature, culture, and ecology, but without the political implications that ecology now entails. Key actors learned their professions within the Soviet educational system and some, like Samaev, Syrenov, Angarkav, and probably others, learned their organizing skills as members of Komsomol. The skills and relationships they brought to the Tunka pipeline debate had been mobilized in environmental initiatives before.

Conclusions: Pipelines, Power, Protest

How do the natural gas and oil pipelines differ? The answer is not simple. On the one hand, both pipelines would bring irreversible damage to people's livelihoods and lifestyles. Given the performance record of Transneft, Yukos, and other Russian energy companies, both would also degrade the environment (Yazovskaya 2006; Balzer 2006; Stammler and Wilson 2006). On the other hand, there continues to be some support for a natural gas pipeline because of its potential to reduce both timber harvesting and air pollution. RUSIA Petroleum currently holds the rights to develop the natural gas reserves at the Kovykta Field in Irkutsk Oblast, and discussion about routing the pipeline through Tunka continues. In a 2005 interview, Nina Krakhmal, acting director of Tunka National Park, expressed concern that any kind of pipeline would permanently damage Tunka's fragile mountain ecosystem and that such damage would offset the ecological benefits of natural gas in reducing atmospheric pollution. Krakhmal went on to describe the catch-22 created by the proposed pipelines through Tunka. Should initial construction result in reversible damage to the ecosystem and it were to recover to a "satisfactory" level, state officials would feel justified in proceeding further—as indeed they would in the case of *ir*reversible environmental damage because, of course, the damage would already have been done. Krakhmal's analysis points to an important aspect of the pipeline development, the critical role played by power relations between Tunka residents, the state, and private industry in achieving social and environmental justice.

As is increasingly recognized in the environmental justice literature, power relations are a principal reason why "dirty" projects generally end up in a less empowered group's backyard (Toffolon-Weiss and Roberts 2005; Pellow and Brulle 2005). With the southern oil pipeline route ruled out, the state is in the process of approving a northern route, which would cut through the herding lands of the Evenki commune (*obshchina*) in Irkutsk Oblast and Buriatiia (Fondahl and Sirina 2006). Shortly after this chapter was completed, the natural gas project again emerged as a threat to the Tunka National Park and the livelihoods of its residents. Two major differences this time are the apparent success of an oil pipeline being built by Transneft along a slightly more northern route and the possibility that the federal government will be fully responsible for the natural gas pipeline through Tunka. In short, Siberia's gas and oil resources seem poised for development.

In the post-Soviet era, citizens of the former Soviet Union have redefined their relationship to the state. Some continue to regard the state as a protector that has an obligation to protect the social and economic welfare of its citizens (Haney 2002; Caldwell 2004). Others feel they have been betrayed by the state, as in the case of Chernobyl, where citizens have sought to redefine their relationship to the state by quantifying the damage caused by this betrayal (Petryna 2002). In Tunka, the creation of the park itself was largely an effort to protect the resources that people use as part of their traditional way of life, yet the identical discourse was used by opponents of the park, who felt the federal status would be too constraining (Metzo 2009).

In local and international protests over the pipelines, the discourse of opposition has centered on the connections between conservation and traditional livelihoods. Environmental justice comes through demanding the state protect the integrity of the park as much for the residents as for any conservation goals. It also comes when the interests of the economically poor, politically marginalized residents of the national park effectively confront the powerful political and economic interests of an energy corporation with close ties to the state. Although efforts of local residents to halt pipelines are significant, the government's role in taking over the development of natural resources is just as significant. TNK-BP is a Russian joint-stock company (with the Tyumen Oil Company [TNK, after Tyumenskaia Neftianania Kompaniia] and British Petroleum [BP] as its principal partners) and the largest shareholder in RUSIA Petroleum. Russian officials have threatened to revoke TNK-BP's license to develop the Kovykta gas field. It is expected that, after May 2007, the state-run energy company Gazprom will take over development of the project (Belton 2007).

Soviet-era relationships to the state were built upon its overwhelming control of every aspect of citizens' lives (Buck-Morss 2000). And even though they were taken care of from cradle to grave, many feel they were not taken care of particularly well. The federal government's inefficiencies have left the population disillusioned with the state. When the natural gas pipeline proposal was believed to be a government initiative, people felt it would remain at the level of talk, hence the apathetic response of Tunka residents when the issue was raised. With the introduction of a second pipeline and the involvement of private businesses, however, both pipelines seemed far more likely to be built, prompting those with access to information about the pipelines to disseminate it as widely as possible to spur public involvement.

"There are only particular, competing, fragmented, and heterogeneous conceptions of and discourses about justice that arise out of the particular situations of those involved," David Harvey (2000, 342) reminds us. Indeed, the debate over pipelines involves a myriad of actors, each with a different version of justice. Mikhail Khodorkovsky's arrest represents one oligarch being held accountable for the advantages he accrued under a weak state. It represents political retribution. And, most important to the residents of Tunka, it represents the elimination of a threat to their way of life. The discourse of traditional livelihoods is also fragmented and used in widely differing ways: as promoting conservation or antithetical to it, as a barrier to the introduction of cleaner technology or as something to be pitied by representatives of energy companies. Even the presentation of this discourse varies greatly, ranging from open-forum letters and articles in newspapers to civil disobedience in public hearings to lawsuits to shamanic rituals. Tunka's successful resistance was not the result of a single, unified organization or movement. Religious actors, though linked to NGOs, operated independently; the local newspaper maintained an impartial, however active, role in disseminating information across social strata; even when working collaboratively, NGOs were not equally involved in the protests. Though itself contestable and temporary, Tunka's success is testimony to the power of effective civil society. By drawing on heterogeneous conceptions and tactics—mobilizing legal expertise, enlisting local venues for discussion, and effectively framing local concerns for those in power, civil society can achieve environmental justice, at least for a time.

Acknowledgments

Research for this chapter was supported in part by IREX and Wenner Gren in 2000–2001 and through a Junior Faculty Research Grant from the University of North Carolina–Charlotte in 2005. Earlier drafts were presented at the 2005 American Anthropological Association meetings in Washington, D.C., George Mason University, and the 2006 Society for Applied Anthropology meetings in Vancouver, B.C. A big thanks to Susan Crate for taking the lead on organizing these events and encouraging me to do this project. Thanks to John Bodley for being a discussant in Vancouver. Revisions were completed during a Reassignment of Duties from the University of North Carolina–Charlotte in 2006. The author alone is responsible for the content, including mistakes.

Notes

1. Ironically, Khodorkovsky was moved to a prison camp in Chita, which was also on the Yukos-proposed pipeline route.

2. The license to develop the Kovykta Field in Irkutsk Oblast, source of natural gas and oil for the two pipelines, belongs to RUSIA Petroleum, an open joint-stock company, whose stockholders include the Irkutsk Oblast government, British Petroleum (BP), and several Russian energy developers. RUSIA Petroleum did not hold the development rights in 2000, when the pipeline was originally proposed.

3. The wide range of issues provides a broader base for environmental activism, yet makes it more complicated to piece together a cohesive movement (Agyeman 2005); it is often the case that specific life events lead to environmental activism (Kempton and Holland 2003, Holland 2003). In trying to identify the extent to

which a movement exists, researchers can act as the glue that begins to bind together a broader movement (see, for example, Babcock 1997; Gillogly and Pinsker 2000).

4. As the locals do, I will use "Tunka" as shorthand for the Tunka Valley, Tunka National Park, and Tunkinskii Raion, except where geography, conservation status, or political organization is being discussed.

5. There was some confusion in the initial discussion about where the eastern terminus of the pipeline would be—Mongolia, China, or the Pacific—because the Russian Federation had not yet identified a partner for the project. I have found no documentation that suggests that Tunka was ever considered a terminus or that the initial plans included servicing the region with natural gas.

6. It is clear from the text of her translated article that Olga Belskaya (n.d.) is addressing a local audience, though the Internet posting makes no note of the article's Russian title or original place of publication.

7. For organizations like Global Response, the explicit goal is to promote the awareness of individuals in the West.

8. According to Belskaya (n.d.), the Kyren public hearings lasted from 10 a.m. to 11 p.m.

9. These monks are all affiliated with the Buddhist ecological organization Akhalar (see table 5.1).

10. For a U.S. parallel to the Ukrainian case, see Kempton and Holland 2003 and Holland 2003.

References

Agyeman, Julian. 2005. *Sustainable Communities and the Challenge of Environmental Justice*. New York: New York University Press.

Anderson, Benedict. 1983. *Imagined Communities: Reflections on the Origin and Spread of Nationalism*. London: Verso.

Babcock, Elizabeth C. 1997. "The EPA and Environmentalism in Chicago: Recommendations for a Community Based Approach." Final report to Society for Applied Anthropology/U.S. Environmental Protection Agency Intern Program. http://www.sfaa.net/eap/eappapers.html/.

Balzer, Marjorie Mandelstam. 2006. "The Tension between Might and Rights: Siberians and Energy Developers in Post-Socialist Binds." *Europe-Asia Studies* 58 (4): 567–588.

Belton, Catherine. 2007. "Russia Threatens to Revoke BP's Siberian License." *Financial Times-Europe*, March 17–18, 4.

Belskaya, Olga. n.d. "Reflections on the Verge of a Disaster or Why I am against Oil Pipelines." Direct Democracy Web site: http://democracy.mkolar.org/Siberia/Reflections.html/ (accessed February 17, 2004).

Brechin, Steven R., Peter R. Wilshusen, Crystal L. Fortwangler, and Patrick C. West. 2003. *Contested Nature: Promoting International Biodiversity with Social Justice in the Twenty-first Century*. Albany: State University of New York Press.

Buck-Morss, Susan. 2000. *Dreamworld and Catastrophe: The Passing of Mass Utopia in East and West*. Cambridge, MA: MIT Press.

Caldwell, Melissa. 2004. *Not by Bread Alone: Social Support in the New Russia*. Berkeley: University of California Press.

Edelman, Marc. 2001. "Social Movements: Changing Paradigms and Forms of Politics." *Annual Review of Anthropology* 30: 285–317.

Evans, Alfred B., Jr. 2006. "Civil Society in the Soviet Union?" In Laura Henry, Evans, and Lisa McIntosh Sundstrom, eds., *Russian Civil Society: A Critical Assessment*, 28–54. Armonk, NY: M.E. Sharpe.

Fondahl, Gail, and Anna Sirina. 2006. "Rights and Risks: Evenki Concerns Regarding the Proposed Eastern Siberia–Pacific Ocean Pipeline." *Sibirica*. 5 (2): 115–138.

Foucault, Michel. 2000. "Governmentality." In James D. Faubion, ed., *The Essential Works of Foucault, 1954–1984*, vol. 3: *Power*, 201–222. New York: New Press.

Fortwangler, Crystal L. 2003. "The Winding Road: Incorporating Social Justice and Human Rights into Protected Area Policies." In Steven R. Brechin et al., eds., *Contested Nature: Promoting International Biodiversity with Social Justice in the Twenty-first Century*, 25–40. Albany: State University of New York Press.

Gillogly, Kathleen A., and Eve C. Pinsker. 2000. "Networks and Fragmentation among Community Environmental Groups of Southeast Chicago." Final report to Society for Applied Anthropology/U.S. Environmental Protection Agency Intern Program. http://www.sfaa.net/eap/eappapers.html/.

Global Response. 2002. "Protect Tunkinskii National Park/Russia." http://www.globalresonse.org/ (accessed March 20, 2006).

Goldman, Marshall I. 2004. "Putin and the Oligarchs." *Foreign Affairs*, November–December, 33–44.

Gupta, Akhil, and James Ferguson. 1997. "Culture, Power, Place: Ethnography at the End of an Era." In Gupta and Ferguson, eds., *Culture, Power, Place: Explorations in Critical Anthropology*, 1–29. Durham, NC: Duke University Press.

Hall, John A. 1995. "In Search of Civil Society." In Hall, ed., *Civil Society: Theory, History, Comparison*, 1–31. Cambridge: Polity Press.

Haney, Lynne. 2002. *Inventing the Needy: Gender and the Politics of Welfare in Hungary*. Berkeley: University of California Press.

Hann, Chris. 1995. "Philosophers' Models on the Carpathian Lowlands." In John A. Hall, ed., *Civil Society: Theory, History, Comparison*, 158–182. Cambridge: Polity Press.

Hann, Chris. 1996. "Introduction: Political Society and Civil Anthropology." In Hann and Elizabeth Dunn, eds., *Civil Society: Challenging Western Models*, 1–26. London: Routledge.

Harvey, David. 2000. *Justice, Nature, and the Geography of Difference*. Malden, MA: Blackwell.

Hengesbach, Alice, and Sergei Shapkhaev. 2003. "Lake Baikal—More Valuable than Oil." *Give and Take*, Autumn: 19–20.

Henry, Laura A. 2006. "Russian Environmentalists and Civil Society." In Henry, Alfred B. Evans, and Lisa McIntosh Sundstrom, eds., *Russian Civil Society: A Critical Assessment*, 211–228. Armonk, NY: M.E. Sharpe.

Holland, Dorothy. 2003. "Multiple Identities in Practice: On the Dilemmas of Being a Hunter and an Environmentalist in the USA," ed. Toon van Meijl and Henk Driesson. Special issue, *Focaal* 42: 31–50.

Igoe, Jim. 2004. *Conservation and Globalization: A Study of National Parks and Indigenous Communities from East Africa to South Dakota*. Belmont, CA: Thompson / Wadsworth Learning.

Johnston, Barbara Rose. 2001. "Anthropology and Environmental Justice: Analysts, Advocates, Mediators, and Troublemakers." In Carole L. Crumley, A. Elizabeth van Deventer, and Joseph J. Fletcher, eds., *New Directions in Anthropology and Environment*, 132–149. Walnut Creek, CA: AltaMira Press.

Johnston, Barbara Rose, ed. 1997. *Life and Death Matters*. Walnut Creek, CA: AltaMira Press.

Kempton, W., and Dorothy Holland. 2003. "Identity and Sustained Environmental Practice." In LaReine Warden Clayton and Susan Opotow, eds., *Identity and the Natural Environment: Intersections*, 317–341. Cambridge, MA: MIT Press.

Ledeneva, Alena V. 1998. *Russia's Economy of Favours: Blat, Networking and Informal Exchange*. Cambridge, UK: Cambridge University Press.

Metzo, Katherine. n.d. "Collaboration on a Tourism Model for Tunka National Park, Russia." Unpublished manuscript.

Metzo, Katherine. 2009. "The Formation of Tunka National Park: Revitalization and Autonomy in Late Socialism." *Slavic Review*, Spring 2009.

Metzo, Katherine. 2001. "Adapting Capitalism: Household Plots, Forest Resources, and Moonlighting in Post-Soviet Siberia." *Geojournal* 55 (2–4): 549–556.

Nineto, Amy. 2005. "'An Island of Socialism in a Capitalist Country': Postsocialist Russian Science and the Culture of the State." *Ethnos*, 70 (4): 443–464.

Palmer, Paula. 2005. "The Pen is Mightier than the Sword: Global Environmental Justice One Letter at a Time." In David Naguib Pellow and Robert J. Brulle, eds., *Power, Justice and the Environment: A Critical Appraisal of the Environmental Justice Movement*, 265–275. Cambridge, MA: MIT Press.

Pazderka, Josef. 2005. "Russia: End of a Messy Affair." *Transitions Online* 118 (June 1). http://www.tol.cz/ (accessed: June 2, 2005).

Pellow, David Naguib, and Robert J. Brulle. 2005. "Power, Justice, and the Environment: Toward Critical Environmental Justice Studies." In Pellow and

Brulle, eds., *Power, Justice and the Environment: A Critical Appraisal of the Environmental Justice Movement*, 1–19. Cambridge, MA: MIT Press.

Peña, Devon G. 1992. "The 'Brown' and the 'Green': Chicanos and Environmental Politics in the Upper Rio Grande." *Capitalism, Nature, Socialism* 3: 79–103.

Peña, Devon G. 2005a. "Autonomy, Equity, and Environmental Justice." In David Naguib Pellow and Robert J. Brulle, eds., *Power, Justice and the Environment: A Critical Appraisal of the Environmental Justice Movement*, 131–151. Cambridge, MA: MIT Press.

Peña, Devon G. 2005b. *Mexican Americans and the Environment: Tierra y Vida*. Tucson: University of Arizona Press.

Pesmen, Dale. 2000. *Russia and Soul*. Ithaca, NY: Cornell University Press.

Petryna, Adriana. 2002. *Life Exposed: Biological Citizens after Chernobyl*. Princeton, NJ: Princeton University Press.

Phillips, Sarah D. 2005. "Civil Society and Healing: Theorizing Women's Social Activism in Post-Soviet Ukraine." *Ethnos*, 70 (4): 489–514.

Samojlova, Nastia. 2005. "Khodorkovskogo posadili v tiur'mu dukhi Tunkinskoi doliny [The Spirits of TunkaValley Put Khodorkovsky Behind Bars]." *Nomer-odin* 15 (April 20): 8.

Sutton, Jennie. 2003. "Lake Baikal and the Thirst for Oil. Report of Pipeline Protest Activities in 2002." http://www.baikalwave.eu.org/Oldsitebew/oil.html/ (accessed January 15, 2007).

Stammler, Florian, and Emma Wilson. 2006. "Dialogue for Development: An Exploration of Relations between Oil and Gas Companies, Communities, and the State." *Sibirica*. 5 (2): 1–42.

Toffolon-Weiss, Melissa, and Timmons Roberts. 2005. "Who Wins, Who Loses? Understanding Outcomes of Environmental Injustice Struggles." In David Naguib Pellow and Robert J. Brulle, eds., *Power, Justice and the Environment: A Critical Appraisal of the Environmental Justice Movement*, 77–98. Cambridge, MA: MIT Press.

White, Anne. 2006. "Is Civil Society Stronger in Small Towns?" In Laura Henry, Alfred B. Evans, and Lisa McIntosh Sundstrom, eds., *Russian Civil Society: A Critical Assessment*, 284–302. Armonk, NY: M.E. Sharpe.

Yazovskaya, Julia. 2006. "The Far East: Pipeline Dreams." *Transitions Online* 173 (June 29). http://www.tol.cz/ (accessed: July 6, 2006).

Yurchak, Alexei. 2003. "Soviet Hegemony of Form: Everything was Forever, Until It Was No More." *Comparative Studies in Society and History* 45 (3): 480–510.

Zhukovskaia, Natalia. 2004. "Neoshamanism in Buriatiia." In Liubov L. Abaeva and Zhukovskaia, eds., *Buriats*, 390–396. Moscow: Nauka.

6

The Role of Culture and Nationalism in Latvian Environmentalism and the Implications for Environmental Justice

Tamara Steger

The environmental movement for independence in Latvia gained momentum in the 1980s as the policy of glasnost was tested and the Russian leadership of the Soviet Union weakened. It facilitated the political events that culminated in Latvian independence and the introduction of a multiparty parliamentary system. This chapter considers the nationalistic and cultural elements of the Latvian environmental movement during these political changes and in early transition. It asserts that at the heart of this movement was a demand for environmental justice.

Latvian environmental justice took shape as the collective demand for protection of nature and the environment came to coincide with the pursuit of national and cultural recognition. It gained momentum as the Soviet Communist system was linked to environmental degradation; "cultural tools" were mobilized; and a discourse on environmental protection was introduced.

Nationalism and Culture in the Latvian Environmental Movement

Animated by nationalism and cultural heritage, environmental activists in Latvia set about democratizing their country and achieving its independence from the highly centralized, authoritarian regime of the former Soviet Union in 1991. As one Latvian environmental and independence movement leader described the time, "Culture exploded!"[1] In this context, environmentalism ultimately created a public forum for expressing collective Latvian opposition to the Soviet regime during the political changes of the late 1980s.

In her study on environmentalism in Russia, Rachel May (1998) concluded that the importance of the environment is integrally tied to national identity and cultural heritage. Katarina Eckerberg (1994, 468) chose different words to make the same point about the Baltic states: "Feelings for nature are deeply embedded in the national heritage." And Jane Dawson (1996) noted that, in several of the former Soviet republics, the quality of the natural environment is a point of national pride. This link between environmentalism, identity, and culture is characteristic of environmental justice movements not only in the former Soviet Union but also throughout the world (Schlosberg 2004; Bullard 2005).

Linking the Soviet System to Environmental Degradation

The Latvian environmental movement succeeded in facilitating the breakdown of the Communist regime, in large part, by politicizing environmentalism (i.e., by linking environmental degradation to the political system). The Latvian environmental movement's struggles against the proposal to build the Daugavpils Dam and hydropower station in a culturally sensitive and nationally symbolic place turned into a protest against the Soviet Communist system itself, which was weakening economically even as it was opening up politically.

For the first time, Latvian environmentalists began to target the system for causing and promoting environmental destruction (hence destruction of the homeland) through its industrial projects (e.g., dams). The dam proposal and the subsequent environmental impact assessments allowed the framers or claim makers (largely members of the intelligentsia) to build a sense of "us" versus "them" along nationalistic lines. When asked how they were able to do this, one movement leader from the University of Latvia replied:

Very simple. We [Latvians] were occupied countries and our environment was destroyed very much [and also] our mental environment. [I]t's something very important for us to defend our environment especially in Soviet Union system since this total Russification.... And, going from region to region destroying environment, polluting by all possible means, Russians who are completely ... I will not say in front of microphone.... So that's why it exploded in this direction and why it was possible to go in this direction.[2]

The nationalistic—and anti-Soviet—framing of environmental issues was critical to arousing popular support for the movement. One Latvian independence/environmental activist reflected on the important role of national symbols in motivating participation in making activists feel that they could, "do something more than just cleanup and restoration activities. That we can engage in political types of demonstrations with our logo and Latvian national colors. It gave us the real feeling that we can do something."

Inherent in this attack on the Soviet system was a critique of the Communist ideology, with its heavy emphasis on economic growth through industry and industrialization, driven by scientific and technical progress and controlled by the state. Bernd Baumgartl (1997, 49) noted the importance of industrial imagery in the Communist ideology:

The symbolic image of working machines and productive factories ensuring the well-being of the population, smoking chimneys which represented the pride of a quickly modernizing society (not occasionally these chimneys often decorated the banknotes and picture postcards of socialist countries), imposed a genuine problem in people's mentality, closely linked to the ruling ideology and political theory—the Communist-Marxist emphasis on economics, industrialization, and technical progress.

Despite the goal of economic growth, inefficiency was rampant within the Soviet system. Energy consumption was high, exceedingly wasteful, and not curtailed in any institutionalized way to improve efficiency (Baumgartl 1997). Centrally established target plans for production bore no relation to any demand structure. Meeting target plans was more important than meeting particular needs, and no responsibility was taken for the outcome.

The Soviet Communist system became the underlying root problem that explained the terrible state of the environment in Latvia. The myth that centralized planning and social ownership promoted economic growth and averted serious environmental problems was exposed. The highly centralized system was now called "irrational" and "inefficient."

For Latvians, the Daugavpils Dam proposal came to symbolize the Communist and Soviet way. Scientific reports and publications on the detrimental ecological impacts of the dam and the lack of economic benefits proved the irrationality and inefficiency of the Soviet Communist system.

Environmental damage and economic inefficiency were not the only or even the most important objects of anti-Soviet sentiment, however. Social injustice stood at the fore. Almost half a century of Russification, of Soviet domination and oppression, during which countless Latvians were sent to prison or deported to Siberia for their political views, created a seedbed of cultural and nationalistic resentment.

The Nazi-Soviet Nonaggression Pact of August 23, 1939 (also known as the "Molotov-Ribbentrop Pact") brought Latvia and Estonia under the control of the Soviets. By 1941, the Latvian government had been abolished and replaced with a "People's" government, and over 14,000 of Latvia's leading citizens had been summarily deported to Siberia (Dreifields 1996; Plakans 1995). One of the protests leading to Latvia's independence more than fifty years later commemorated the dates of this pact and the deportations.

The 1950s in Latvia were marked by a massive in-migration of Russians. As part of the policy of Russification, and to meet factory employment needs, almost a million "Russian-speaking settlers" were relocated to Latvia after World War II.[3] Before the war, Latvians had made up 75 percent of the total population; by the late 1980s, they made up only 50 percent.[4] Housing complexes were erected for incoming Russian workers. Daugavpils, a former industrial city some 120 kilometers (75 miles) from the Russian border, became over 50 percent Russian as a result of the in-migration of Russian workers after World War II.

When Latvia officially gained independence in 1991, it readopted the 1922 Latvian Constitution. Under a law enacted by the new republic, to qualify for Latvian citizenship, applicants had to pass a Latvian language and history test and be residing in Latvia before 1940 or be born of parents who resided in Latvia before 1940. The law posed a formidable challenge for Latvia's minority residents: only one out of ten could speak and write Latvian (Clemens 1998). Under a second law, all Russians, whether they came willingly or unwillingly to Latvia after World War II, had to apply for Latvian citizenship. Responding to criticism from European human rights groups (Clemens 1998) and from the European Union, Latvia eventually moderated its citizenship laws.

The Russian language was increasingly marginalized. In 2000, the main newspaper in Latvia, *Diena*, stopped printing its Russian edition; environmental non-governmental organizations started publishing in Latvian or English, even though there were many residents who could speak and read Russian.[5]

Once independence was achieved, ethnic Russians who had supported Latvian independence were disillusioned by their political exclusion. These Russians "had felt that they would be supported by Latvians because they supported [their] independence," a Latvian cultural historian and environmental activist recalled, "but they were a little bit wrong."

Mobilizing Cultural Tools

In addition to nationalistic sentiment targeted against the Soviet-based centralized regime, culture created a collective context for action and helped to define the tools available to activists. Cultural recognition is central in the struggle for environmental justice (Schlosberg 2004); Latvia's environmental and independence movement, which reached its peak in the late 1980s, was notably infused with Baltic cultural elements, especially singing. Ann Swindler (1986) observed that culture can be a "tool kit" for action. And, indeed, singing, pagan rituals, and nature poetry (as in the Daina) have served as the tools in how environmental activism has evolved in Latvia.

Environmental activities, especially those performed by students, merged with cultural work; indeed, their main purpose was cultural. "Students were working not just eight hours, but some days up to sixteen hours and through days and nights," recalled one of the movement's leaders, "cleaning up national parks or building nature trails or restoring fisheries, cultural monuments, castles, etc. These students were preserving our cultural heritage, doing real practical work for restoring the environment, nature, and culture. The student environmental movement started with this restorative work."

Student participation in the environmental movement was steeped in cultural symbolism. A Latvian student described what she did: "I was going throughout Latvia and taking care of the landscape. For example, particularly involving work that is national Latvian symbol such as cleaning up oak trees overgrown by different bushes. It is most important tree mentioned in all our legends and Daina." One insightful visitor to the country said it was as if the students "were laboring against the Soviet Union by restoring their cultural and natural heritage."

Before the regime change, more or less secret meetings were held at the University of Latvia, in which Latvian culture and history were the main themes of the lectures. A prominent environmental activist at the time and a specialist in Latvian culture (subsequently employed by the Latvian Ministry of Culture and the Latvian Cultural Museum) had gained access to special Latvian historical documents, whose contents were distributed during these meetings.

Singing is part of the Latvian sense of identity—"Singing since birth, singing as I grow; a life spent singing" goes the refrain to one of the best-known Daina. Indeed, Latvia is a singing culture, a tradition dating back nearly 1,000 years. As one journalist noted, "Latvians sang through conquests by the Swedes, Germans, Russians, and the Soviets" (Lyons 2003). Latvian folk songs, called "Daina," came from ancient poetry that was collected and compiled by Krisjanis Barons back in the 1920s. In his travels across Latvia, Barons gathered 218,000 song texts that were subsequently set by musicians to 30,000 melodies. In addition to being about love and war, many are about nature. Baltic singing festivals convene about 30,000 singers and dancers. Although Latvians were allowed to have their singing festivals during Soviet times, they were not allowed to sing certain nationalistic songs such as "God Bless Latvia," and the Soviets attempted to convert the events into Soviet propaganda.

The revolution that brought Latvian independence is often referred to as the "Singing Revolution" (Thomson 1992). A Latvian in the independence movement explained: "From the point of view of not just physically, but mentally to keep our identity, our language, to use nonviolent method of resistance or whatever you call it. We did this through our singing festivals and all other kinds of ways, hiding in words of songs. That's why it was called a 'Singing Revolution.'"

When the Latvian Supreme Council adopted a resolution to restore Latvian independence in May 1990 (independence became official in 1991), people took to the streets and marched with the justice officials on their shoulders to the Daugava River. Two weeks after this event, however, there was a confrontation with the Russian opposition in Latvia known as "Interpol." This group appeared before the Latvian Parliament building in direct opposition to the Latvian independence organizers. The interaction then took an especially interesting turn. In a research interview I conducted, a Latvian environmental activist described this turn:

Latvian Environmental Activist: Interpol officers [the Russian Front] started to attack the parliamentary building. The Latvians formed a chain and there was a confrontation, but by singing fortunately.
Steger: By what?
Latvian Environmental Activist: By song. We made interference by song. They sang special Russian songs, memorials of the Second World War, and we sang Latvian folk songs. We did it all the time at demonstrations, made a chain and started to sing. We had no weaponry at all.

Introducing a Discourse on Environmental Protection

The Latvian environmental movement merged with the independence movement linking environmental degradation and Soviet rule; in this union, nature, nationalism and culture were asserted. As the movement evolved, the notion of environmental protection was introduced—which served as the launching pad for a public forum on the "politics of pollution" (Bullard 2005). Together with nationalism and culture, it created a spark for environmental justice in Latvia.

A shift from the politically safer notion of nature protection to the politically charged concept of environmental protection was integral to the role of the environmental movement in the larger political changes at work in Latvia. Close examination of this shift also reveals emerging social categories for marking and marginalizing different individuals and groups as the regime was changing. The terms of political inclusion were being redefined as issues of environmental justice, rights, and equity were articulated.

Nature protection was in many ways associated with the Soviet Communist regime, whereas environmental protection represented the new political frontier. Carried out largely by volunteers, nature protection activities focused on protecting particular areas such as parks, monuments, and riparian zones, through cleanup campaigns, small ecological restorative efforts (e.g., establishing buffers around rivers), and educational programs. The educational programs, in turn, emphasized biology, entomology, and geology, among other natural science fields, and were aimed at teaching volunteers what they needed to be able to answer questions such as "I know that this is a butterfly, but what kind?"

Nature protection did not address environmental issues most directly linked to the industrial activity at the heart of the Communist ideology—pollution. Instead, it emphasized the appreciation of nature for ecological reasons rather than, say, for reasons of human health. Based on information in the natural sciences, nature protection excluded social science issues that would raise questions about politics, economics, quality of life—and ultimately justice. The health and welfare of the individual were not addressed. Educating people about butterflies, for example, posed little, if any, threat to the regime.

The notion of environmental protection, on the other hand, provided a "completely new point of view," as one early Latvian environmentalist and university professor remarked. In addition to the adoption of this "new point of view," new ways of doing environmental activist work were subsequently introduced. Large amounts of funding for environmental NGOs came from western Europe and the United States, and were earmarked for training activists, to build the professional skills they needed to run organizations. Some of these activists received funding to go to the United States for such training. Funding institutions further encouraged professionalism by requiring plans, annual reports, and measurable results.

Western-funded and -organized trainings and workshops emphasized "capacity building": fundraising skills, strategic planning, grant application procedures and approaches, report writing, meeting organization, and outreach skills. International organizations also conducted team-building and group dynamic exercises, and Latvian environmentalists began to join others across the globe in the critique of industrialization and consumerism.

This global critique was largely concerned with the effects of industrialism on people (Lipschutz 1996). Environmental protection accordingly focused on reducing pollution by influencing national policies or conducting large-scale projects. Whereas environmental groups before the regime change provided ecological education in biology and entomology, the Children's Green School in newly independent Latvia stressed environmental protection issues such as waste management in their educational programs.

Environmental protection involved such activities as policy work, lobbying, report writing, and administrative tasks including record keeping and fundraising. The field became increasingly dominated by paid professionals working alongside volunteers. Whereas nature protection, for the most part, took place locally and on site (such as at a park or on a river), environmental protection had national, international, and global implications. Indeed, advocates of environmental protection were mostly members of the political opposition during the push for independence.

Environmental protection addressed the issue of pollution, which was linked to industries governed by the Soviet state. Not only did it challenge the fundamental ideology of the Communist system by questioning the industrial mode of production, it publicized a "politics of pollution." Emphasizing environmental problem solving, environmental protection called for action to reduce pollution, improve water quality, and reduce waste. Its larger, people-oriented view of ecology included economics, politics, and society. Through education about the impacts of particular industrial activities or projects on the environment, human health, and the economy, people were given a more universal rationale for questioning the state, one that went beyond nationalism and culture. The affirmation of the inevitable relationship between environmental quality and human equality was at the core of this emerging discourse (see Agyeman et al. 2002).

Implications for Environmental Justice in Transition

Three principal factors emerge in considering the implications for environmental justice in Latvia in transition: defining the perpetrators of injustice, locating responsibility for the environment, and considering the role of diverse streams of environmentalism.

Although Latvians in the environmental movement moved vigorously to challenge the Soviet regime on issues of environmental justice, they became less inclined in transition to challenge their own government on issues of environmental justice in order to preserve their country's independence and stability. Environmental NGOs in post-Soviet Latvia, for example, were hesitant to point the finger at the Latvian state as the cause or source of an environmental problem. Post-independence protests tended to target "outsiders" such as foreign countries and foreign-owned firms. For example, environmentalists protested against an oil spill at a Lithuanian Oil Terminal that polluted the Baltic Sea in 2001 by marching on the Lithuanian Embassy in Riga. In another such protest, environmentalists campaigned against a foreign-owned mobile telephone company proposing to build a station in a culturally sensitive area.

Once the Latvian state had been established and environmental protection efforts were increasingly institutionalized, environmental NGOs' notion of primary responsibility for protecting the environment shifted largely to the individual. Indeed, for some environmentalists in Latvia, commitment to environmental change was a question of "being right in your soul." One Latvian leader of a prominent environmental organization explained that the reason for "environmental pollution is not that there are good or bad laws or good or bad ministers of environmental protection or irresponsible institutions, the reason is that people somehow don't care and [that reason is] in their soul." Another

Latvian environmental NGO representative asserted that even organized environmental activities were less important in addressing environmental problems than individual changes: "It's important to understand that we cannot solve environmental problems with a few actions and activities and so on. First of all we have to solve the problems in ourselves."

After the political changes, however, Latvian environmental NGOs did engage environmental justice issues at a more procedural level in the form of general initiatives to promote public participation (e.g., Aarhus Convention) and local involvement in decision making. Furthermore, the urban, internationally connected, professionalized environmental NGOs are the primary actors on environmental justice issues at this universal level. For example, the Riga-based environmental organization Green Liberty has worked to promote the Aarhus Convention on Access to Information, Public Participation in Decision Making and Access to Justice in Environmental Matters; the Latvian Fund for Nature has been involved in public policy making to improve public participation.

At its grassroots volunteer level, the Latvian environmental movement has focused on issues such as nature restoration and cleanup campaigns (REC 1994). At its more professionalized and institutionalized level, the movement has adopted the environmental protection approach, chiefly addressing pollution problems in a wider policy context (Steger 2004). Bridging these two streams of environmentalism in Latvia lays the foundation for a discourse on just sustainability by uniting different frameworks for action (Agyeman 2005). In Latvia, such a discourse could mutually benefit both the professionalized and the more local and rural environmental NGOs in the promotion of environmental justice.

Conclusion

The Latvian environmental movement for independence was a call for environmental justice. The articulation of anti-Soviet sentiment, the mobilization of cultural symbols and tools, and the introduction of a discourse on environmental protection allowed a "politics of pollution" to emerge. It brought to the forefront issues of environmental justice, rights, and equity. In Latvia's transition, the struggle for environmental justice faces a new set of challenges and circumstances markedly different from those of the Soviet era.

Notes

1. Interviewees in this research were promised that their names would not be used in any publications. Hence brief descriptors are offered for all informants quoted, but their precise identity is not revealed.

2. Most of the interviews conducted for this research were conducted in English or with an informal Latvian translator (usually arranged by the interviewee, and frequently someone from within the interviewee's organization). The interviews were directly transcribed from audiotapes without editing.

3. Juris Dreifields (1996) is careful about how he refers to people and their nationality. In this case, for example, he refers to the people coming into Latvia as "Russian-speaking settlers" rather than as "Russians." As will be discussed at greater length later, the issue of language is prominent in the debate on Latvian citizenship. To qualify for Latvian citizenship, applicants must be able to speak Latvian. Many of the older people who came to Latvia from Russia back in the 1950s and never learned Latvian are hard pressed to gain citizenship because of the language requirement.

4. These population statistics are taken from http://www.latinst.lv/history.htm/.

5. Language was a principal means by which Latvians could reassert their national identity over ethnic Russians residing in the country.

References

Agyeman, Julian. 2005. *Sustainable Communities and the Challenge of Environmental Justice*. New York and London: New York University Press.

Agyeman, Julian, Robert Bullard, and Bob Evans. 2002. "Exploring the Nexus: Bringing Together Sustainability, Environmental Justice, and Equity." *Science and Polity*, 6 (1): 77–90.

Baumgartl, Bernd. 1997. *Transition and Sustainability: Actors and Interests in Eastern Europe Environmental Policies*. London: Kluwer Law International.

Bullard, Robert D. 2005. *The Quest for Environmental Justice: Human Rights and the Politics of Pollution*. San Francisco: Sierra Club Books.

Clemens, Walter C. 1998. "The Baltic Reborn: Challenges of Transition." *Demokratizatsiya*, 6 (4): 710.

Dawson, Jane. 1996. *Eco-Nationalism*. Durham, NC: Duke University Press.

Dreifields, Juris. 1996. *Latvia in Transition*. Cambridge: Cambridge University Press.

Eckerberg, Katarina. 1994. "Environmental Problems and Policy Options in the Baltic States: Learning from the West?" *Environmental Politics*, 3 (3): 445–478.

Lipschutz, Ronnie D. 1996. *Global Civil Society and Global Environmental Governance*. Albany: State University of New York Press.

Lyons, Michael. 2003. "Keeping in Tune with the Land that Sings." *Baltic Times*, July 14.

May, Rachel. 1998. "Environmental Nationalism and Russia's Conservation Movement: Ideals of Nature and the National Parks." Project Report for the National Council for Soviet and East European Research.

Plakans, Andrejs. 1995. *The Latvians: A Short History*. Stanford, CA: Hoover Institution Press, Stanford University.

Regional Environmental Center (REC). 1994. *Public Participation Manual*. Szentendre, Hungary. http://www.rec.org/REC/Publications/PPManual_Baltic/Latvia.html/ (accessed January 14, 2007).

Schlosberg, David. 2004. "Reconceiving Environmental Justice: Global Movements and Political Theories." *Environmental Politics*, 13 (3): 517–540.

Steger, Tamara. 2004. "Environmentalism in Latvia and Hungary: Turning from East to West." In Charlotte Wallin and Daniel Silander, eds., *Democracy and Culture in the Transatlantic World*. Växjö, Sweden: Växjö University Press.

Swindler, Ann. 1986. "Culture in Action: Symbols and Strategies." *American Sociological Review* 51 (2): 273–286.

Thomson, Clare. 1992. *The Singing Revolution*. London: Michael Joseph.

7

The Fight for Community Justice against Big Oil in the Caspian Region: The Case of Berezovka, Kazakhstan

Kate Watters

After the breakup of the Soviet Union, the new Republic of Kazakhstan wasted no time inviting Western corporations to invest in its economy. Seeing its vast natural resources as the key to economic development, and Western investment as an alternative to the historical economic dominance of Russia, Kazakhstan began negotiating with transnational oil companies in the mid-1990s, inviting them to develop Tengiz and Karachaganak, two of its most lucrative onshore oil and gas fields. The offshore Kashagan Field was discovered in the mid-1990s and was quickly dubbed the "largest oil find of the past twenty years." Western oil companies scrambled for deals with the Kazakhstani government and soon began to develop the fields; since then, revenue has flowed steadily into government and corporate coffers, even as local citizens closest to the fields continue to live in dire poverty.

The other former Soviet states of Turkmenistan, Azerbaijan, and Russia also continue to struggle with the economic difficulties that have plagued them since the dissolution of the Soviet empire. Following the model of many authoritarian, centralized governments around the world, these states have focused on oil wealth to the exclusion of other economic avenues (Karl 1997) and have sacrificed their poorest and most vulnerable citizens to the economic and environmental injustices of the oil boom.

Many factors complicate the picture, including the absence of an agreed-upon legal status of the Caspian Sea, disputes over ownership of offshore oil blocks, and the need to rebuild environmental protection systems, which were lost when the sea suddenly found itself bordered by five rather than two countries.

Environmental Concerns and Oil Development

Western corporations and state oil companies are engaged in extensive oil and gas extraction from on- and offshore petroleum fields in the Caspian region, which has resulted in a variety of environmental concerns. These include not only global problems such as climate change, but also immediate economic hardship and environmental destruction in the communities where international oil companies are active. The Caspian Sea is home to many endemic species, from the beluga sturgeon to the Caspian seal, which are registered on the World Conservation Union (IUCN) Red List as endangered (IUCN 2008). These and other animals are at serious risk from habitat degradation and loss. Inadequate oil spill response systems; lack of clarity about environmental regulations; and corruption, graft, and lack of resources also threaten the region's environment.

This chapter addresses the even more egregious human cost of oil and gas development. From lost agricultural jobs and environmental health problems in refinery communities to the massive hiring of local residents into temporary construction jobs in the oil industry and the increased incidence of sexually transmitted diseases in traditional communities close to transient worker camps (Bacheva et al. 2006), petroleum production has intensified a downward spiral of unsustainable economic development and ecological degradation not only in Kazakhstan but in the wider Caspian region as well. Much of this activity is bankrolled by international finance institutions such as the World Bank and the European Bank for Reconstruction and Development (EBRD), resulting in U.S. and European taxpayer funding of environmentally harmful, economically unsound development.[1] In other words, the very institutions whose mandate is to alleviate poverty around the world are financing the impoverishment of the region's most vulnerable populations in rural, poverty-stricken regions beyond the capital cities.

U.S. companies play an enormous role in this exploitation: Chevron is the largest private oil investor in Kazakhstan (Chevron 2008). ExxonMobil has a major share in the development of the offshore Kashagan Field, which threatens seal whelping and sturgeon spawning grounds. Subsidiaries of Halliburton have joined Chevron and ExxonMobil in the Caspian Pipeline Consortium (CPC) project, despoiling miles of steppe en route to the Black Sea port of Novorossiysk, Russia, as it transports oil from the Tengiz and Karachaganak oil fields to consumers in the West. Just as the major oil companies have exploited low-income communities in the United States (Ryder 2006), they have done so around the world (Agyeman et al. 2003), including in the Caspian region.

Western corporations such as Halliburton, Unocal, and ExxonMobil are active even in repressive Turkmenistan, where the government routinely imprisons civil society and environmental activists, and where citizens are deprived of the most basic human rights.[2] Dragon Oil has developed the offshore Cheleken oil field, and Western companies continue to attempt to negotiate a trans-Caspian pipeline that would bypass both Russia and Iran, bringing Turkmenistan's reportedly substantial natural gas reserves to the West via Azerbaijan. Since the December 2006 death of former President Saparmurat Niyazov, Western oil companies are tripping over each other for a share of Turkmenistan's natural gas reserves.

In Azerbaijan, Western oil companies are actively involved in the major oil fields; Azeri-Chirag-Guneshli, one of the largest offshore oil fields, is dominated by BP. The 1,700-kilometer (1,050-mile) Baku-Tbilisi-Ceyhan (BTC) Pipeline, which transports Azeri oil to the Turkish port of Ceyhan, has triggered enormous environmental and social resistance in the region as it threatens numerous protected territories and agricultural regions of Azerbaijan, Georgia, and Turkey. Human rights abuses along the pipeline are rampant.[3]

The story of oil and gas development in the Caspian region corresponds to that of many other resource-rich nations around the world. Suffering from the "resource curse," the nations of the Caspian Sea region provide a vivid case study of the serious shortcomings of oil and gas development for local economies, and the pervasive nature of the injustice of resource extraction (Tsalik 2003).

This chapter focuses on the impact of the Karachaganak oil and gas condensate field on the surrounding communities, and the local response to environmental degradation, health problems, and human rights abuses that have resulted from the development of the field.

Located in northwestern Kazakhstan, the Karachaganak Field comes within 20 kilometers (12 miles) of the Russian border (figure 7.1). The field is one of the largest oil and gas condensate fields in the world and was discovered in 1979 by the Soviets, who began its exploration and initial development. With highly sulfurous gas condensate, the field presents significant technical challenges, and it was not until the late

Figure 7.1
Map of Kazakhstan.

1990s, when the international consortium, Karachaganak Integrated Operating, B.V. (now Karachaganak Petroleum Operating, B.V., or KPO), invested in the project, that the field was developed to its capacity.

In 1997, Karachaganak Petroleum Operating signed a production-sharing agreement (PSA) with the Kazakhstani government, under which KPO will operate the Karachaganak Field until 2038 (see http:// www.kpo.kz). The consortium comprises British Gas, ENI/Agip (Italy), Chevron (United States), and LUKoil (Russia), with 32.5 percent shares each for BG and ENI, 20 percent for Chevron and 15 percent for LUKoil. KPO not only extracts oil and gas from the field, but also transports Karachaganak oil to Orenburg, Russia, for refining; it has built a new pipeline with access to the CPC pipeline. Second only to Tengiz in production, Karachaganak provides enormous revenues both to the consortium and to the Republic of Kazakhstan.

The Karachaganak Field covers an area of over 280 square kilometers (70,000 acres or 110 square miles); it holds more than 1,200 million metric tons of oil and condensate and over 1.35 trillion cubic meters (47.75 trillion cubic feet) of gas. KPO's expansion of the field has involved an investment of over $4.3 billion, making Karachaganak the biggest internationally funded project in Kazakhstan (see http:// www.kpo.kz).

In 2002, the International Finance Corporation (IFC), the private lending arm of the World Bank, provided KPO with $150 million in loans.

Although the IFC had first conducted an environmental impact assessment, it failed to take into consideration the impact of the field on the nearby village of Berezovka and other surrounding communities.[4]

Berezovka: The Case for Environmental Justice

Berezovka, Kazakhstan is a village of 1,300 residents, which is located 5 kilometers (3 miles) from the Karachaganak oil and gas condensate field. A former state-run collective farm (sovkhoz) during the Soviet period, the village once housed large herds of cattle and a sausage factory. Situated on the sweeping steppes of Kazakhstan, Berezovka is nestled on the shores of the Berezovka River; its residents proudly grow their own vegetables, care for fruit orchards, harvest berries, and tend to their own livestock. Largely self-sufficient, the villagers are accustomed to a life of hard work and, until recently, rich rewards as their harvests yielded food enough for winter, milk from their own cows, jams from their orchards, and the satisfaction of tending their own gardens. A village school and a community center form the heart of the community, where children and adults share fellowship, education, and the social life of the village.

Berezovka is located at the end of a dirt road 25 kilometers (16 miles) from the town of Aksai—the main city in Burlinsky Raion, where Karachaganak Petroleum Operating has its headquarters. It is one of nine villages located around the Karachaganak Field, and the closest to the field ever since the villagers of Tungush, 3 kilometers (2 miles) from Karachaganak and very close to Berezovka, were relocated to the city of Uralsk in 2003. Until then, residents of the two villages shared similar environmental and health problems, and had begun to work together for environmental justice.[5]

As a result of chemical emissions at Karachaganak, Berezovka residents breathe toxic air, drink toxic water, and grow their food in toxic soil. However, Berezovka has joined the global environmental justice movement as a community fighting to preserve its way of life (Bullard 2005). Berezovka activists are fighting the same kinds of injustices that local communities in the Niger Delta (Okonta and Douglas 2001) and on the Gulf Coast (Ryder 2006)—and poor communities around the world—are. Scholars have documented similar cases in South Africa (Bullard 2005) and elsewhere and have pointed to the negative impacts of mining and extraction that have attended globalization (Crate 2006; Karl 1997). Berezovka's struggle is a classic fight for environmental justice, including the right to a clean and healthy environment, which is guaranteed by both the Constitution and the Law on the Environment of the Republic of Kazakhstan.[6]

The Aarhus Convention and Environmental Justice

Berezovka activists and their partners believe that Karachaganak Petroleum Operating has violated the Convention on Access to Information, Public Participation in Decision Making and Access to Justice in Environmental Matters (commonly called the "Aarhus Convention"), to which Kazakhstan is a signatory, by failing to include the public in decision making regarding activities at the Karachaganak Field. The Aarhus Convention guarantees the public the right to be informed of environmentally significant matters, to participate in environmentally significant decision making, and to legal redress in cases when their right to be informed or participate in decision making has been denied.[7] The government of Kazakhstan as well as international corporations and financial institutions operating there are required to comply with the convention. On the international level, the Aarhus Convention operates in much the same way that national right to know laws function—it gives citizens the right to know what pollutants and toxins are present in their community and provides them with a mechanism to access and participate in environmental decision making.

This decision making often involves the placement of oil and gas facilities, pipeline routing, and other matters that will have serious consequences for local communities. Particularly in instances like the Berezovka case, poor communities, communities of color, and those in remote places are the last to be notified of pending environmental decisions, even though they will be the most directly impacted (Agyeman et al. 2003). The Aarhus Convention provides a way for these communities to demand their right to be informed, and, more important, their right to have a say in reaching the decisions that will affect them. Green Salvation, an Almaty-based NGO, which also actively campaigns at Karachaganak, is the first group in Kazakhstan to bring a case before the Aarhus Secretariat for noncompliance with the convention (see http://www.greensalvation.org/en). This precedent has helped the Berezovka Initiative Group use the Aarhus Convention more effectively in its struggles; as they better understand their rights, group members are demanding full participation in environmental decision making.

In March 2008, the Berezovka Initiative Group and Green Salvation won their first major Aarhus Convention victory. The Supreme Court of Kazakhstan ruled that the Western Kazakhstan Department of Statistics was required to release to the public statistical information on atmospheric emissions at Karachaganak that it had withheld as "proprietary" and "confidential." The court found the information to be environmental in nature and therefore subject to public review. Its ruling was all the more significant for being the first time the court referred to the convention as binding in Kazakhstan (having previously considered it no more than advisory).[8]

The Sanitary Protection Zone

According to Kazakhstani law, each potentially dangerous industrial facility operating within the country must be surrounded by a "sanitary protection zone" (SPZ)—an area where no one is allowed to live. The zone serves as a buffer between the industrial facility and the local population, ensuring that no one is exposed to toxins at levels dangerous to human health. Because of the nature of production at the Karachaganak oil and gas condensate field, the minimum legal SPZ for the field is 5 kilometers (3 miles) (Akhmedyarov 2006).

And that was the minimum when Karachaganak Petroleum Operating became the operator. Because the two villages were located within the zone, according to Kazakhstani law, the residents of both Tungush and Berezovka were entitled to be relocated to safe and environmentally clean sites. In May 2002, the Berezovka Initiative Group received a letter from the Ministry of Environment and Natural Resources of Kazakhstan to that effect.[9] The local authorities and KPO announced in the local press that they would build "Villages of the Twenty-First Century" for the residents of Tungush and Berezovka, which lacked indoor plumbing, hot water, and natural gas, most homes being heated by wood-burning stoves (Sokovnin 2003).

This seemingly open-and-shut situation was much complicated in 2003, however. Without prior public notice—let alone public participation—and in clear violation of the villagers' human rights, KPO and the government of Kazakhstan signed an agreement reducing the SPZ to 3 kilometers (2 miles) and effectively excluding Berezovka from the relocation requirement. The reasons given were KPO's "superior technology" and the need to adopt current international standards of operation.[10]

No residents of Berezovka or any of the other villages surrounding the Karachaganak Field participated in this decision or were even aware the matter was being discussed. According to the terms of the Aarhus Convention, the Kazakhstani government was obliged not only to inform the public about the planned reduction of the sanitary protection zone, but also to invite residents of Berezovka and other impacted villages to participate in deciding whether to proceed with it. As a company operating in Kazakhstan, KPO was also obliged to comply with the convention.

Working for Environmental Justice: Community-Based Activism

Led by music teacher Svetlana Anosova, members of the Berezovka Initiative Group, formed in 2002, began to connect their worrisome health symptoms with the constant gas flaring, strong smell of sulfur, and increased industrial activity at Karachaganak. The villagers noticed that they and their children had developed respiratory illnesses, skin rashes, vision problems, nosebleeds, and other ailments. Virtually all the villagers noticed problems with their gardens and livestock. Although, in the beginning, the villagers simply sought to identify the source of these problems, when it became clear that the Karachaganak Field was to blame, they began to demand that their community be relocated to a safe and environmentally clean site where they could raise their children without fear of toxic exposure. They did not come lightly to the conclusion that resettlement was their best option, but as long as the Karachaganak facility was spewing toxins into the air, they saw no other option.[11]

The Berezovka Initiative Group first talked with their village and regional mayors (*akims*), and even with members of Parliament, hoping that, once it became clear they were suffering health problems as a result of the development of the Karachaganak Field, the authorities would curtail the toxic emissions and improve the situation in their community. They also appealed to the Health Safety and Environment officers at KPO, requesting that they conduct more intensive environmental monitoring in Berezovka to determine the extent of the environmental burden the community was bearing.

It soon became clear, however, that neither the authorities nor Karachaganak Petroleum Operating had any interest in helping Berezovka solve its health problems.[12] KPO stated that, according to its production-sharing agreement, the health of the villagers was the responsibility of the local authorities, and the local authorities, using data provided by KPO, stated that there were no environmental problems from the Karachaganak Field. Instead, they suggested that the air pollution problems in Berezovka were the result of burning family trash piles.[13]

When the villagers realized they would have to fight for relocation, the Berezovka Initiative Group asked Crude Accountability to work with them, to provide partnership and training, and to help them build a strategy for their relocation campaign.[14] And so the villagers began to compile their own health and environmental data, using a community-based methodology developed in toxic communities in the United States. This data collection methodology and the villagers' subsequent advocacy have served as the basis for their environmental justice campaign.

Starting in 2003, the Berezovka Initiative Group collaborated with Crude Accountability and Kazakhstani environmental NGOs Green Salvation (Almaty) and TAN (Atyrau) to identify patterns of illness in the village, monitor toxic emissions from Karachaganak, observe environmental changes, and officially bring concerns and information requests to the Kazakhstani government, the World Bank, and KPO. They also steadily articulated concrete demands for compensation and resettlement, and informed the public about the environmental hazards they were being exposed to every day. They have developed an arsenal of tools in their struggle to defend their rights, and engaged experts when necessary. They have learned how to work with the media, and they have cooperated with those journalists sympathetic to their cause. They have successfully carried out letter-writing campaigns to the ministries and the presidential apparatus, informing the national government in Astana, the capital, of their plight. And, by attracting national and international attention, the Berezovka Initiative Group has effectively mobilized support for their cause.

Proving Their Case: The Villagers Undertake Scientific Monitoring, Analysis, and Advocacy

Health Monitoring

The government of Kazakhstan conducted medical research into the health of the Berezovka residents in 2002, and again in the spring of 2004. Dr. U. I. Kenesariev conducted extensive medical research, not only in Berezovka, but also in other villages close to the Karachaganak Field. The results of his research, although made available to Karachaganak

Petroleum Operating, were never released in full to the villagers. In fact, when asked for the results of the study in August 2004, he explained that they were "proprietary." Because KPO had paid for them, he was unable to give the information to anyone else.[15] However, this study is used repeatedly by KPO to support their claim that the health situation in Berezovka is no worse than in other communities farther away from Karachaganak and therefore, according to KPO, not adversely impacted by operations at the field.

Similarly, a 2004 health study of Berezovka by the regional health authorities was never fully explained to the villagers. Repeated requests for the information have gone largely unanswered, even though the authorities frequently refer to the study when talking about the environmental health situation at Karachaganak.

Fed up with the unresponsiveness of KPO and the Kazakhstani Ministry of Health, the villagers began their own health study in 2003. Using "popular epidemiology," a grassroots methodology deployed in toxic communities in the United States for the past fifteen years, and provided to the village by U.S. environmentalist Linda King (2004; see also Novotny 1998) and Crude Accountability, from December 2002 to May 2003, Anosova and her colleagues went door to door, asking villagers to describe their symptoms.[16] After talking with members of approximately 400 households, representing the 1,286 residents of Berezovka, they discovered that 45 percent of the village population was suffering from chronic medical problems. Children, in particular, suffered from toxic exposure to the Karachaganak Field, experiencing epilepsy, dizziness, skin ailments, nosebleeds, aggressivity, memory loss, and vision problems; adults, for their part, suffered from debilitating nausea, skin rashes, vision problems, and memory loss. Their findings are displayed in tables 7.1 and 7.2.

According to the villagers, virtually every woman of childbearing age was suffering from some disruption of her menstrual cycle, and the incidence of stillbirths and birth defects is still on the rise. Most of this information is anecdotal in nature, and it has been difficult to confirm the incidence of illness among women, in particular. Cultural norms make discussion of cancer, "female problems," and childbirth taboo. Only after working in the community for four years did Crude Accountability begin to learn more about these problems.[17]

The villagers' analysis of the health situation in Berezovka included a mapping exercise, which demonstrated that the highest incidence of ill-

Table 7.1
Adult health data, Berezovka, Kazakhstan, 2002–2003

Health problems among adult residents	Residents with symptom(s) (N = 886)	
	Number	Percentage
Loss of memory	688	79
Muscular-skeletal problems	599	69
Significant loss of hair	423	49
Loss of teeth	423	49
Loss of vision	413	47
Cardiovascular problems	401	46
Gastroenterological problems	375	43
Respiratory problems	308	35
Skin ailments	260	30

Table 7.2
Youth health data, Berezovka, Kazakhstan, 2002–2003

Health problems among youth	Middle school students (N = 80)		High school students (N = 100)	
	Number	Percentage	Number	Percentage
Overall weakness	38	47.5	95	95
Severe headaches	45	56.3	83	83
Loss of memory	24	30	77	77
Frequent fainting spells	20	25	77	77
Skin ailments	29	36.3	67	67
Aggressivity[a]	—	—	49	49
Nose bleeds	28	35	34	34
Chest pains[b]	21	26.3	—	—

[a] Middle school students were not asked this question.
[b] High school students were not asked this question.

ness occurred in low-lying areas and along the banks of the Berezovka River. For example, in the northwestern corner of the village, closest to Karachaganak and along the riverbank, households reported a high incidence of skin, respiratory, circulatory, and nervous system problems.

Finally, in December 2004, the villagers hired the services of an independent health clinic to conduct a sample blood analysis of Berezovka residents. Approximately sixty villagers volunteered to participate to determine whether they had been exposed to toxins.

The villagers, most of them women and children, traveled by car from Berezovka to Aksai to have their blood drawn. In addition, they answered an oral questionnaire given by Crude Accountability staff. The results of the blood analysis and questionnaire clearly demonstrated that the majority of those tested were suffering from chronic, often serious, health problems that could be connected to the toxins coming from Karachaganak.

The blood work was analyzed not only by the independent clinic but also by Aleksey Lazko, a doctor based in Astrakhan, Russia. In his report, Dr. Lazko stated that the results of the analysis, showing high levels of anemia, low white blood cell counts, and other factors, were consistent with toxic exposure to hydrogen sulfide associated with petroleum production (Lazko and Lazko 2005).[18]

Environmental Monitoring

In addition to the health monitoring, the Berezovka Initiative Group conducted independent environmental monitoring, focusing primarily on airborne toxins. They also monitored the village's drinking water, cow's milk from a resident's livestock, and snowfall.

The villagers were trained in the "Bucket Brigade" air-monitoring method by certified Crude Accountability trainers, who traveled to Berezovka in August 2004.[19] A simple, inexpensive grassroots technique employed in oil-impacted communities throughout the United States and the world, the Bucket Brigade uses a 5-gallon bucket, valves, vacuums, sterile bags, and tubing to create an effective air-monitoring device. By simulating the human lung, the bucket collects a 3-minute air sample, using a method approved by the United States Environmental Protection Agency. Bucket Brigade air samples are then analyzed in an independent laboratory that can detect the presence of toxic chemicals.[20]

The villagers took air samples with buckets over the course of a year. Whereas the local authorities and KPO sampled for four to six toxic chemicals,[21] the Bucket Brigade sampling collected over twenty-five toxic substances, which were found to include carbon disulfide, methylene chloride, toluene, hydrogen sulfide, and other toxins—among them, substances the local authorities were not even testing for (table 7.3).

When the toxic emissions from Karachaganak blow toward Berezovka, the air frequently smells of rotten eggs or has a sickly sweet odor that sticks to the back of your throat. Headaches, confusion, and nausea accompany these smells, and villagers know that KPO is emitting toxins into the air. On mornings when emission levels are high, a navy blue fog hangs in the low-lying areas of the village, particularly on the north side of the community. Village elders, many of whom live in this neighborhood, complain of nausea and of disturbances among their livestock when these conditions are reported.

Armed with buckets, the villagers conduct their air sampling during these hours. Often, they sample in the middle of the night or the early morning, when KPO most often spews the highest levels of its emissions. Before Bucket Brigade sampling, KPO and the Ministry of Ecology dismissed the villagers' claims that the smells from the field were making them sick. The authorities claimed that the smell of hydrogen sulfide did not indicate elevated emissions, and that the villagers' discomfort was not associated with toxic exposure.

The Bucket Brigade sampling showed otherwise: when the smells were strongest, the emission levels were dangerously high. The villagers' common sense and local and historical knowledge of their environment were, in fact, triggering alarms that would protect the community. However, it was a matter of cataloging and organizing that information that would make the difference with the authorities. Turning local knowledge, kitchen talk, and anecdote into powerful information was the most important work of the Berezovka Initiative Group (table 7.4).

Empowered by the impact of the air-monitoring results, the Berezovka Initiative Group also collected samples in December 2004 from the water intake into Anosova's house. Crude Accountability transported these samples to an independent laboratory in Orenburg, Russia, where they were tested for heavy metals, other toxins, and radioactivity. Although the water tested negative for radioactive contamination, it was deemed "not of drinking water quality" by the laboratory, which found that heavy metals were within the acceptable range, but that there were high levels of chlorides (CSSEIOO 2004).[22] These findings were shared with Berezovka residents and distributed widely.

Table 7.3
Chemical compounds identified in Berezovka Bucket Brigade air monitoring 2004–2005 above "levels of concern" (Crude Accountability)

Chemical compound	Date	Level in μg/m³	US level of concern in μg/m³	Level in mg/m³	Kazakhstani maximum permissible concentration for daily exposure mg/m³	Kazakhstani maximum permissible concentration for acute exposure (mg/m³)	Health effects of chronic exposure	Health effects of acute exposure	EPA cancer classification
Methylene chloride	9/11/2004	16	4.09	0.016	0.01	0.01	Headaches, dizziness, nausea and memory loss. Animal studies indicate effects on the liver, kidney, central nervous system and cardiovascular system.	At high levels can be fatal; effects on the nervous system include: decreased visual, auditory, and psychomotor functions (reversible once exposure ceases)	Group B2, probable human carcinogen
	8/20/2005	5.5		.0055					
	8/24/2005	5.9		.0059					
	8/25/2005	5.9		.0059					
	8/29/2005	8.8		.0088					
Carbon disulfide	9/11/2004	13	3	.013	0.005	0.03	Muscle pain, headaches and general fatigue	Vomiting, nausea, dizziness, fatigue, headache, mood changes, lethargy, blurred vision	Not classified for carcinogenic effects
	12/1/2004	6.5		.0065					
	7/20/2005	10.8		.0108					
	8/20/2005	7.8		.0078					
	8/29/2005	9.3		.0093					
Toluene	9/11/2004	170	95.8	.17	—	0.6	Significant decreases in lung function, asthma-like reaction characterized by wheezing, dyspnea, and bronchial constriction	Severe irritation of skin and eyes, affects respiratory, gastrointestinal and central nervous systems	Group B2, possible human carcinogen. (Animal studies have reported significantly increased incidences of tumors of the pancreas, liver and mammary glands from exposure)
	12/1/2004	10		.01					
	7/20/2005	12		.026					
	8/20/2005	26		.019					
	8/24/2005	19		.02					
	8/25/2005	21		.034					
	8/29/2005	34							

Acrylonitrile	9/11/04	20	0.028	0.02	0.03	—	Headaches, fatigue, nausea, weakness	Mucous membrane irritation, headaches, dizziness and nausea	Probable human carcinogen
Hydrogen sulfide	9/11/2004 7/20/2005	5.09 7.29	1	0.0059 .00729	—	0.008	Hydrogen sulfide does not accumulate in the body. But repeated or prolonged exposure has been reported to cause low blood pressure, headache, nausea, loss of appetite, weight loss, ataxia, eye membrane inflammation, and chronic cough. Neurologic symptoms, including psychological disorders, have been associated with chronic exposure. Chronic exposure may be more serious for children.	Nausea, headaches, delirium, disturbed equilibrium, tremors, convulsions and skin and eye irritation. Inhalation of high concentrations can produce extremely rapid unconsciousness and death.	Not classified for carcinogenic effects
Acetone	9/11/2004 7/20/2005 8/20/2005 8/29/2005	65 31 29 39	—	.065 .031 .029 .039	0.15	0.05	Long-term exposure to acetone has resulted in kidney, liver and nerve damage, increased birth defects and lowered ability for males to reproduce in animals. Moderate levels of acetone result in respiratory irritation and burning eyes in humans.	Breathing moderate-to-high levels of acetone for short periods of time can cause nose, throat, lung and eye irritation; headaches, light-headedness; confusion; increased pulse rate; effects on blood; nausea; vomiting; unconsciousness and possibly coma; and shortening of the menstrual cycle in women.	Not a known carcinogen

Table 7.3
(continued)

Chemical compound	Date	Level in μg/m³		US level of concern in μg/m³	Level in mg/m³		Kazakhstani maximum permissible concentration for daily exposure mg/m³	Kazakhstani maximum permissible concentration for acute exposure (mg/m³)	Health effects of chronic exposure	Health effects of acute exposure	EPA cancer classification
Xylenes (m, p and o)		m, p	o	—	m, p	o	0.006	0.2	Chronic exposure of humans to mixed xylenes, as seen in occupational settings, has resulted primarily in neurological effects such as headaches, dizziness, fatigue, tremors, in coordination, anxiety, impaired short-term memory, and inability to concentrate. Labored breathing, impaired pulmonary function, increased heart palpitation, severe chest pain, abnormal EKG and possible effects on the kidneys have also been reported. Mixed xylenes	Acute inhalation exposure to mixed xylenes in humans has been associated with dyspnea and irritation of the nose and throat; gastrointestinal effects such as nausea, vomiting, and gastric discomfort; mild transient eye irritation; and neurological effects such as impaired short-term memory, impaired reaction time, performance decrements in numerical ability, and alterations in equilibrium and body balance. Acute animal studies have reported	No information available
	9/11/2004	6.9			.0069						
	7/20/2005	5.1			.0051						
	8/20/2005	16	6.5		.016	.0065					
	8/24/2005	13	5.4		.013	.0054					
	8/25/2005	18	7.5		.018	.0075					
	8/29/2005	21	8.5		.021	.0085					

respiratory, cardio-
vascular, CNS, liver
and kidney effects from
inhalation exposure to
mixed xylenes.

have been shown to
produce developmental
effects, such as an
increased incidence of
skeletal variations in
fetus, delayed ossifica-
tion, fetal resorptions,
and decreased fetal
body weight in animals
via inhalation expo-
sure. Some studies
observed maternal
toxicity as well.

US EPA studies show
that repeat exposure
adversely affects the
reproductive system
and developing fetus of
animals.

Breathing large
amounts of 1,2,4-
Trimethylbenzene for
short periods of time
adversely affects the
human nervous system.
Effects range from
headaches to fatigue
and drowsiness. The
vapor irritates the nose
and throat. The liquid
will irritate the skin.

1,2,4-Trimethyl-benzene + Unidentified Siloxane	8/24/2005 8/25/2005	10 20	6.2	.01 .02	0.15	0.04

Table 7.4
Crude Accountability Bucket Brigade air monitoring sample results 2004–2005

Compound	μg/m³	US level of concern in μg/m³	How many times sample exceeds the US level of concern	mg/m³	Kazakhstani maximum acute permissible concentration mg/m³	How many times sample exceeds acute Kazakhstani MPC	Kazakhstani average daily maximum permissible concentration mg/m³	How many times sample exceeds average daily Kazakhstani MPC	Total toxins registered
9/11/2004									
Hydrogen sulfide	5.09	1	5.09	0.0059	0.0080	0.74	—		
Carbon disulfide	13.3	3	4.33	0.0130	0.03	0.43	0.0050	2.60	
Acrylonitrile	20	0.028	714.29	0.02	—		0.03	0.67	
Toluene	170	95.8	1.77	0.17	0.60	0.28	—		
Methylene chloride	16	4.09	4.00	0.0160	0.01	1.60	0.01	3.20	
Acetone	65	370		0.0650	0.05	1.30	0.15	0.43	
2-Butanone (MEK)	27								
m,p Xylenes	6.9								
Propane	70								
Isobutane	80								
n-Butane	50								
Ethanol	200								

Compound								
2-Methylbutane	200							
Isopropyl alcohol	30							
n-Pentane	100							
2-Methylpentane	20							
3-Methylpentane	20							
n-Hexane	20							
Methylcyclopentane	20							
1-Butanol	40							
3-Methylhexane	20							
n-Octane	20							22
12/1/2004								
Carbon disulfide	6.5	3	**2.16**	0.0065	0.03	0.22	0.0050	
Toluene	10	95.8	0.10	0.01	0.06	0.17	—	**1.30**
Chlorodifluoromethane + Propene + Propane	20							
Ethanol	100							
Isobutane + acetone	40							
2-Methylpentane	70							
Cyclohexane	20							
2,3-Dimethylheptane	20							
Octyl acetate	20							
C13H28 branched alkane	90							
C15H32 branched alkane	90							

Table 7.4
(continued)

Compound	μg/m³	US level of concern in μg/m³	How many times sample exceeds the US level of concern	mg/m³	Kazakhstani maximum acute permissible concentration mg/m³	How many times sample exceeds acute Kazakhstani MPC	Kazakhstani average daily maximum permissible concentration mg/m³	How many times sample exceeds average daily Kazakhstani MPC	Total toxins registered
C15H32 branched alkane	20								
C15H32 branched alkane	30								
C15H32 branched alkane	20								13
7/20/2005									
Hydrogen sulfide	7.27	1	**7.27**	0.0073	0.0080	0.9088	—		
Carbon disulfide	10.8	3	**3.60**	0.0108	0.0300	0.3600	0.0050	**2.1600**	
Acetone	31	370	0.08	0.031	0.065	0.4769	0.15	0.2067	
Toluene	12	95.8	0.13	0.01	0.06	0.2000	—		
m,p xylenes	5.1			0.0051	0.20	0.0255	0.0060	0.8500	
Ethanol	200								
n-Decane	20								7

8/20/2005								
Methylene chloride	5.5	4.09	1.3447	0.0055	0.0100	0.5500	0.0050	1.1000
Carbon disulfide	7.8	3	2.6000	0.0078	0.0300	0.2600	0.0050	1.5600
m,p-Xylenes	16			0.0160	0.2000	0.0800	0.0060	2.6667
o-Xylene	6.5			0.0065	0.2000	0.0325	0.0060	1.0833
Acetone	29	370		0.029	0.500	0.0580	0.15	0.1933
Toluene	26	95.8		0.026	0.60	0.0433	—	
Propane + carbonyl sulfide	3							
n-Butane	3							
Ethanol	9							
n-Decane	4							
2-Ethyl-1-hexanol	3							
C12H24 compound	4							12
8/24/2005								
1,2,4-Trimethylbenzene + unidentified siloxane	10	6.2	1.61	0.01	0.04	0.25	0.015	0.67
Methylene chloride	5.9	4.09	1.44	0.0059	0.0100	0.5900	0.0050	1.1800
m,p-Xylenes	13			0.013	0.20	0.07	0.0060	2.17
o-Xylene	5.4			0.0054	0.20	0.03	0.0060	0.90
Toluene	19	95.8	0.20	0.019	0.07	0.29	—	
n-butane	30							
Ethanol	70							

Table 7.4
(continued)

Compound	µg/m³	US level of concern in µg/m³	How many times sample exceeds the US level of concern	mg/m³	Kazakhstani maximum acute permissible concentration mg/m³	How many times sample exceeds acute Kazakhstani MPC	Kazakhstani average daily maximum permissible concentration mg/m³	How many times sample exceeds average daily Kazakhstani MPC	Total toxins registered
Isopentane	40								
n-Pentane	20								
2-Methylpentane	20								
2-Ethyl-1-hexanol	20								
n-Undecane	10								
C12H24 compound	30								13
8/25/2005									
Methylene chloride	5.9	4.09	**1.44**	0.0059	0.0100	0.5900	0.0050	**1.1800**	
1,2,4-Trimethylbenzene + unidentified siloxane	20	6.2	**3.23**	0.0200	0.0400	0.5000	0.0150	**1.3333**	
Toluene	21	95.8	0.22	0.0210	0.06	0.3500	—		
m,p-Xylenes	18			0.0180	0.20	0.0900	0.0060	**3.0000**	
o-Xylene	7.5			0.0075	0.20	0.0375	0.0060	**1.2500**	

Ethanol	70					0.07	5.00	0.0140
Isopentane	20					0.02	4.00	0.0050
Hexamethylcyclotrisiloxane	10							
2-Ethyl-1-hexanol	20							
C14H18 branched alkane	10							
C12H24 compound	20							[11]
8/29/2005								
Methylene chloride	8.8	4.09	**2.15**	0.0088	0.0100	0.8800	0.0050	**1.7600**
Carbon disulfide	9.3	3	**3.10**	0.0093	0.0300	0.3100	0.0050	**1.8600**
Toluene	34	95.8	0.35	0.0340	0.0600	0.5667	—	
m,p-Xylenes	21			0.0210	0.2000	0.1050	0.0060	**3.5000**
o-Xylene	8.5			0.0085	0.2000	0.0425	0.0060	**1.4167**
Ethanol	90							
Isopropyl alcohol	10							
n-Undecane	20							
1-Octanol	60							
Acetone	39							
2-Butanone (MEK)	7							[11]

Sources: Laboratory Analysis from Columbia Analytical Services, Inc.; Resources for Science and Justice http://www .creativelement.com/~gwen/locform2.html, "Utverzhdeny prikazom U.O. Ministra zdravookhraneniz respubliki Kazakhstan "18" August 2004 no. 629 "Ob utverzhdenii sanitarno epidemiologicheskikh pravil i norm po epidemiologii i gigene" 2004 18 Sentia-bria no. 3076."

Once the residents realized that their "feelings" about what was happening in their village were backed up by facts, the Berezovka Initiative Group began to enjoy a larger membership, and the community became more outspoken in its criticisms of Karachaganak Petroleum Operating. This "people's research" identified critical problems within the community. Official attempts to discount the knowledge of the villagers became more difficult once the community began to organize.

Advocacy and Mobilization for Relocation

Armed with the environmental data they had gathered, the Berezovka Initiative Group began to meet with human rights activists and others who they thought might help them communicate more effectively with the local authorities. In meetings with the Western Kazakhstan Oblast Ecology Department in Uralsk, for example, it had become clear to group members that, although the environmental officials were familiar with maximum permissible concentrations of chemicals, they were not familiar with environmental law, the Aarhus Convention, or the Kazakhstani Constitution. The villagers decided to educate themselves about these in order to better inform government officials of their responsibilities to the citizens of the community.

In December 2004, the Berezovka Initiative Group and Crude Accountability, in partnership with Russian environmental lawyer Olga Yakovleva, held a human rights workshop in Berezovka to examine the rights expressly described in the Constitution of the Republic of Kazakhstan. Close to one hundred members of the Berezovka community attended: the workshop hall was filled to overflowing, with people packed in at the back of the room and standing in the hallway outside. Attempts by the local authorities to intimidate the villagers into not attending backfired when elderly women scolded police officers for behaving badly, and journalists from nearby Aksai and Uralsk scrambled for places in the room.

Following the workshop, Berezovka residents sent a letter to President Nursultan Nazarbaev, requesting that he intervene in the ongoing violations of human rights in Berezovka, including the villagers' right to a clean environment, right to assemble, and right to take care of their own health. They specifically referred to harassment and intimidation that had occurred in Aksai when they had gathered to give blood samples: police officers attempted to force their women into cars to take them to the police station for questioning. In the following months, the villagers also wrote letters to the senior sanitary doctor of the Republic of Kazakhstan and the Ministry of Environment, asking them to investigate the environmental and health situations in Berezovka.[23]

The Kazakhstani press—much of it controlled either by the government or by Karachaganak Petroleum Operating—ran a series of articles about how Western forces were manipulating the Berezovka villagers into believing they should be relocated by KPO. The articles stated that most Berezovka residents did not want to leave the village and were being misrepresented by a small group of extremists who were under the influence of—and financed by—Crude Accountability (Korina 2005; *Burlinsky Vesti* 2004).

In response to these accusations, the Berezovka Initiative Group and Crude Accountability commissioned a sociological survey in 2005 by researchers from the Russian Academy of Sciences' Institute of Sociology. Residents were asked a series of questions about the environmental conditions in the village, whether they wanted be relocated, and, if so, how. Virtually the entire village participated in the survey: 258 heads of households were interviewed. Survey results showed that the participants overwhelmingly supported relocation for the village: over 90 percent of the village residents wanted to be relocated; over 60 percent of those who were for relocation wanted to be relocated together. Furthermore, over 50 percent of the residents believed that they had no choice but to move because of the village's toxic exposure from Karachaganak. And, among those who opposed relocation, most were elderly pensioners who simply wanted to die in their home village. According to the survey researchers, only ten of the individuals interviewed in the survey wanted to stay in Berezovka (Belov and Khalii 2005).

Engaging the International Community

In August 2004, the Berezovka Initiative Group submitted an official complaint to the Office of the Compliance Advisor Ombudsman (CAO) at the International Finance Corporation (IFC), the private investment arm of the World Bank. The complaint requested an investigation into the IFC's financing of Karachaganak and stated that the activity at the field had negatively impacted Berezovka both economically and environmentally, that the sanitary protection zone had been illegally reduced, and that the development of the field was harming the health of the village residents. It specifically requested that the CAO's office investigate the reasons for reducing the sanitary protection zone from 5 to 3 kilometers (3 to 2 miles). The office accepted the complaint in October 2004 and sent a team to Berezovka in December 2004 to investigate the villagers' claims.

The CAO's initial report, issued in April 2005, stated that, because Karachaganak Petroleum Operating was in compliance with International Finance Corporation requirements, neither KPO nor the IFC had a responsibility to relocate the villagers of Berezovka. However, the report also pointed to the need for KPO and the IFC to establish more effective mechanisms for communication with the villagers. It noted that KPO had failed to operate with adequate transparency, particularly in not explaining either the reason for the reduction of the sanitary protection zone or its environmental monitoring data to the communities around the field. It recommended that KPO establish a method for regular and more comprehensive information sharing with the villagers (CAO 2005).

On May 26, 2005, Crude Accountability and the Berezovka Initiative Group wrote to Karachaganak Petroleum Operating (copying the CAO's office), recommending the two sides collaborate to implement the CAO's suggestions. The group agreed they should meet with KPO, and that Crude Accountability and the local administration should also be in attendance. They recommended that the CAO's office moderate the sessions to ensure that its recommendations were addressed as intended. CAO staff expressed an interest in serving this function if KPO would agree to come to the table.

Karachaganak Petroleum Operating stated that it would meet with the local administration, CAO, and the Berezovka Initiative Group, but only if Svetlana Anosova agreed to come on her own, without Crude Accountability, Green Salvation, or "any NGOs." The CAO's office advised Anosova to accept these conditions, assuring her she would have the opportunity to seek the advice of her colleagues in an unofficial capacity, and explaining that if she did not accept, the office would be unable to facilitate further negotiations between the Berezovka Initiative Group and KPO.

In July 2006, Anosova wrote to the CAO, stating that the time had come and gone for negotiation: in the two years since her group had first filed the complaint, the situation had changed. The Berezovka Initiative

Group would deal directly with the government, rather than with Karachaganak Petroleum Operating.[24] Shortly after receiving Anosova's letter, the ombudsman branch of the CAO closed the case and turned it over to the compliance branch.

In April 2007, the compliance officer of the CAO began to audit the International Finance Corporation's monitoring activity at Karachaganak to determine whether it had failed to meet IFC requirements.[25] On April 25, 2008, the CAO reported that it found "the monitoring program and the data reported on stack emissions [at Karachaganak] insufficient...to verify compliance with IFC requirements." Moreover, "neither the ambient air quality monitoring program nor the data reported from the monitoring to date verify compliance with IFC requirements" (CAO 2008). The report revealed that the IFC failed to report hydrogen sulfide levels from 2003 to 2006, although this toxin is one of the major environmental health dangers to the community surrounding the field.[26] According to IFC requirements, the Karachaganak case will remain open at the CAO, with regular monitoring, until the project comes into compliance.

In a separate set of events, in April 2007, the U.S. Department of Justice found the U.S. company, Baker Hughes, guilty of violating the Foreign Corrupt Practices Act (FCPA), in part for bribing a Kazakhstani official in order to win a lucrative contract at Karachaganak. Baker Hughes paid $44 million in fines—the largest fines levied in the history of the FCPA (see DOJ 2007; SEC 2007). Because the International Finance Corporation was financing Karachaganak Petroleum Operating at that time, Crude Accountability and the Berezovka Initiative Group filed a complaint with the Department of Institutional Integrity (INT) at the World Bank, asking it to investigate the situation, and to bar Baker Hughes, KPO, and Roy Fearnley—the individual employee of Baker Hughes directly involved in the scandal—from receiving any further financing from the World Bank. The complaint was accepted and is under investigation by INT.

Engaging the Kazakhstani Government

The Berezovka Initiative Group's active campaign for relocation and environmental justice has spurred a series of investigations by the Kazakhstani government and elicited public support by the regional deputy to Parliament, Amanzhan Zhamalov. In April 2006, the public prosecutor for Western Kazakhstan Oblast announced to the press that the sanitary protection zone around Karachaganak had been illegally reduced (Akhmedyarov 2006). The increased attention to the problems around Karachaganak, including this statement, resulted in widespread media attention to the demands of the Berezovka villagers and in the creation of a special commission within the Ministry of Health to reexamine the health situation there. The commission traveled to Berezovka in July 2006, ostensibly to study the environmental health situation in the community. In point of fact, it conducted no new research. Instead, as recorded in a letter to the Berezovka Initiative Group and Green Salvation, the Ministry of Health simply recycled previous environmental monitoring data from KPO, which indicated no elevated emissions from the Karachaganak Field.[27]

At a January 22, 2007, public meeting (*skhod*) in the village of Berezovka, Deputy Zhamalov again expressed his concern about the environmental health situation in Berezovka. Local government officials then told those gathered they had allocated an additional 300 million tenge (some 2.5 million dollars) for environmental health monitoring to be conducted in the village in February 2007. Deputy Zhamalov encouraged the villagers to express their demands as to how the monitoring should be conducted.[28]

Then, in October 2007, the Kazakhstani government, acknowledging that the protection zone around Karachaganak had been illegally reduced in size, officially reinstated the 5-kilometer SPZ. The Berezovka Initiative Group and Green Salvation both wrote the government, requesting clarification: did the reinstatement entitle Berezovka to be relocated? In May 2008, the government confirmed the size of the sanitary protection zone as five kilometers, without commenting on the relocation issue.

Media Allies

As the Berezovka Initiative Group has gathered environmental, health, and social data, it has provided this information to the media in Kazakhstan. Although the country's media are subject to significant government control, and, in Burlinsky Raion, to pressure from KPO, a few brave journalists have reported on the environmental and human rights violations at Karachaganak.

One of these publications is the *Uralsk Weekly*, a newspaper published in the regional city of Uralsk, about a two-and-a-half-hour drive northwest of Berezovka. Its coverage of Berezovka's struggle, and KPO's blatant disregard for Kazakhstani law, has played an important role in the village's struggle for justice.

In April 2005, the *Uralsk Weekly* reported that the Western Kazakhstan Oblast Ecology Department had refused to issue an operating license to KPO for 2005 because of numerous environmental violations. The violations included spewing 56,000 metric tons of toxic emissions into the atmosphere, improper storage of solid toxic waste on the Karachaganak Field, and dumping toxic effluent into the water table. Indeed, the consortium had fulfilled only two of the seven benchmark requirements the Oblast Ecology Department had set for it in 2004 (Zlobina 2005).

Uralsk Weekly journalists had previously covered stories from Berezovka, including the harassment of women and children when they went to have their blood drawn for testing in December 2004. They had written about the human rights seminar conducted in the village to educate residents about their environmental rights. Local authorities had attempted to shut down the seminar, surrounding the building with KNB (Kazakhstan's equivalent of the KGB) officers and local police. *Uralsk Weekly* journalists interviewed local residents and gave their harassment by the police prominent coverage (Akhmedyarov 2004). These stories were then picked up by progressive journalists in other parts of Kazakhstan and received national coverage (Dashkov 2004).

The *Uralsk Weekly* continues to publish articles about the situation at Karachaganak, even when threatened with closure.[29] During 2004, *Respublika*, a national newspaper that ran articles about Berezovka, was in fact shut down, causing speculation that other "opposition" newspapers would also be closed. Despite the risk, *Uralsk Weekly* continues to advocate for Berezovka.

Conclusion

The Berezovka Initiative Group has come to understand that what started as a fight for the villagers' right to live in a safe community with clean air, soil, and water has turned into a fight for environmental justice in the broadest, most active sense. The group's campaign for environmental justice has the potential to directly change the lives of 1,300 people for the better. It may also impact others indirectly by setting an important precedent of citizen action in the Caspian region. It sends a message to the oil companies, the World Bank, and the Kazakhstani government that, if they choose to endanger any community, no matter how small or remote, the people will fight back.

In the wake of the Rose, Orange, and Tulip Revolutions, Kazakhstani authorities are nervous about the presence of Western NGO representatives in communities where there is widespread dissatisfaction. Crude Accountability is perceived as aggravating this dissatisfaction in Berezovka; its representatives and Berezovka Initiative Group members have been accused of trying to promote a "Color Revolution" within Kazakhstan (Kenzhegalieva 2006). One newspaper article went so far as to describe this chapter's author and Svetlana Anosova as "agents of Western interest" and even to publish the author's passport data (*Burlinsky Vesti* 2004).

The police and the KNB have repeatedly threatened the villagers. Nor has the intimidation been confined to Berezovka. When Anosova came to Washington, D.C., in the summer of 2003 to meet with officials of the World Bank about the situation in Berezovka, one executive director of the bank told her, "Karachaganak is good for Kazakhstan on the macro level. In this world there are winners and losers and in this situation, you are the losers."

On the national level, there are many questions, chief among them, will the government relocate Berezovka now that the 5-kilometer sanitary protection zone has been reinstated?

Whatever happens, the campaign has already changed the equation of power between the transnational corporations operating in the Caspian region and the local communities and NGOs. Activists are, for the first time, demanding that corporations operate in accordance with international law and environmental best practice. By educating and informing local residents, this campaign places the necessary information in the hands of those who have been considered powerless, enabling them to stand and fight for their basic human rights. With tools like popular epidemiology, the community has been able to challenge the authorities, bringing their concerns to decision makers on the national and international levels.

Relocation for Berezovka will provide a powerful example for the rest of the world. Just because a community is small does not mean it is inconsequential; just because it is in a remote location does not make it powerless; just because it is poor does not mean it is ignorant.

Acknowledgments

Thank you to the members of the Berezovka Initiative Group, who continue to struggle for their environmental rights, and especially to Svetlana Anosova, whose brave, strong and compassionate leadership has sustained the community for the past five years. Thank you to our colleagues at Green Salvation and TAN for their expertise and partnership at Berezovka, as well as to Linda King and Mark Warford, who have provided vision and leadership for Crude Accountability and the Berezovka Initiative Group. Thank you to the Trust for Mutual Understanding, the Underdog Fund of the Tides Foundation, and the Sigrid Rausing Trust for their support of Crude Accountability's and the Berezovka Initiative Group's work on the Karachaganak campaign. Without their generosity, we could not undertake this work. Thank you to Michelle Kinman, who not only skillfully edited the chapter, but has also been engaged in the Karachaganak campaign from the beginning. And, finally, my heartfelt thanks to Jim and Ella Boissonnault, who support every aspect of the Karachaganak campaign. Any errors in this chapter are my full responsibility.

Notes

1. The World Bank and the European Bank for Reconstruction and Development have provided significant funding for petroleum projects in the Caspian region. According to a recent study by the Bank Information Center, funding for oil and gas projects by international financial institutions represents overall 57 percent of their financing. The International Finance Corporation's oil and gas portfolio represents 90 percent of its lending to Azerbaijan, and the EBRD's, 66 percent (Mainhardt-Gibbs 2006).

2. For more information about environmental concerns and oil investment in Turkmenistan, see Crude Accountability 2007.

3. For more information on the environmental and social costs of the Baku-Tbilisi-Ceyhan Pipeline, see the Web sites of the Bank Information Center (http://www.bicusa.org) and the CEE (Central and Eastern European) Bankwatch Network (http://www.bankwatch.org).

4. International Financial Corporation project officer, conversation, International Finance Corporation, Washington, DC, August 2004.

5. The villagers of Tungush were relocated under very controversial circumstances. According to eyewitnesses, they were given one day to pack and were moved from their village to a suburb of Uralsk. Many were left unemployed following the move, and others continued to work on the private farm many of

them worked at in Tungush. This meant that, although the village was deemed uninhabitable because of its proximity to Karachaganak and the exceedences of maximum permissible concentrations of toxins in the air, the men who worked at the farm continued to live in a dormitory in Tungush. The other Tungush residents were moved into a high-rise apartment building just outside Uralsk. The building had been left half built for six years before KPO decided to finish it and pay to have Tungush residents resettle there. Six months following the resettlement, Berezovka Initiative Group members visited the building, which was already beginning to fall apart. When they saw how the residents of Tungush had been relocated, the group created their own set of demands regarding the acceptable terms of resettlement for Berezovka's residents.

6. Article 5 of the Republic of Kazakhstan's Law on the Environment states that citizens, to whom it grants the right to information about the state of the environment, also have an obligation to protect the environment, raise their level of environmental understanding, and work to educate the population about environmental protection. See Zakon Respubliki Kazakhstan ob Okhrane okruzhaiushchei sredi (c izmeneniami v sootvestsvii c Zakonom Respubliki Kazakhstan, May 11, 1999, no. 381-1 Zakonom Respubliki Kazakhstan, November 29, 1999. no. 488-1) [The Law of the Republic of Kazakhstan on Environmental Protection (with changes in accordance with the Law of the Republic of Kazakhstan, May 11, 1999, no 381-1 of the Law of the Republic of Kazakhstan, November 29, 1999, no. 488-1)], Sbornik Zakonodatel'nykh aktov Respubliki Kazakhstan [Collection of Legislative acts of the Republic of Kazakhstan], pp. 3–22.

7. The Aarhus Convention was signed in Aarhus, Denmark, on June 25, 1998, and came into force in both Azerbaijan and Kazakhstan in October 2001. For more information on the convention, see Zaharchenko and Goldenman 2004; http://www.unece.org/env/pp/treatytext.htm/ (accessed March 8, 2008).

8. Green Salvation's announcement of this decision can be found at http://www.greensalvation.org/en/.

9. The Berezovka Initiative Group has a copy of the May 29, 2002 letter from the Ministry of Environment in its files. An English-language translation is available from Crude Accountability upon request.

10. The CAO report found that, in consideration of KPO's "new point-source monitoring technology," the Kazakhstani government allowed the sanitary protection zone around Karachaganak to be reduced. See CAO 2005, 14.

11. Svetlana Anosova, interview by Crude Accountability staff, Berezovka, November 2003, in Crude Accountability 2003.

12. Kazakhstan's political system does not allow for the election of local government representatives. Village, city, county, and regional mayors (*akims*) are appointed by the presidential apparatus. Thus the village mayor is not directly accountable to the citizens of his community, but rather to the city or regional mayor who has appointed him (who has, in turn, been appointed by the regional mayor, who was appointed by the president of Kazakhstan).

13. Svetlana Gumyenaya, resident of Berezovka, interview by Crude Accountability staff, November 2003, in Crude Accountability 2003.

14. At the time, Crude Accountability staff were working for the U.S. nongovernmental organization ISAR, which provided technical assistance and grants to environmental organizations in the former Soviet Union (see http://www.isar.org). In 2003, Crude Accountability was created and began working full-time on environmental justice concerns in the Caspian region, including with the Berezovka Initiative Group.

15. Dr. U. I. Kenesariev, conversation, Uralsk, Kazakhstan, August 2004.

16. Linda Price King, founder and former Executive Director of the Environmental Health Network and author of *Chemical Injury and the Courts, A Litigation Guide for Clients and Their Attorneys* (1999), worked with toxic communities in the United States and around the world for over fifteen years, using popular epidemiology and other community-based techniques to empower local residents in their struggles for environmental justice. The data collected by the Berezovka Initiative Group and Crude Accountability are available at http://www.crudeaccountability.org/.

17. A program specifically directed toward female health problems started up in early 2007, at the suggestion of Svetlana Anosova, head of the Berezovka Initiative Group. The project will address women's health concerns, identifying key problems, their sources, and methods for improving the health of the women villagers.

18. Dr. Aleksey Lazko has worked with toxic communities in Russia for over twenty years and has extensive experience monitoring the exposure of villages close to petroleum-processing facilities close to Astrakhan. Dr. Lazko's report on Berezovka is available from Crude Accountability.

19. Crude Accountability staff were trained and certified as Bucket Brigade monitors and trainers by Global Community Monitor (see http://www.gcm.org/ [accessed January 12, 2007]). With equipment and training manuals, translated into Russian for local use, they conducted several days' training in Berezovka, teaching the local residents how to use the equipment. By the end of the training, they had certified a dozen Berezovka residents in Bucket Brigade monitoring.

20. The Berezovka air samples were tested at the Columbia Analytical Laboratory (CAL) in Simi Valley, California. The Bucket Brigade methodology was developed by Global Community Monitor (see http://www.gcm.org) and has been used in toxic communities throughout the United States and around the world. The Berezovka Initiative Group is the first community group to employ the technology in the former Soviet Union. For more on CAL's findings for the air samples collected in Berezovka in 2004–2005 and on their public health implications, see http://www.crudeaccountability.org/en.

21. Vladimir Khon, testimony at public hearing held in Uralsk, Kazakhstan, May 25, 2006, videotaped by Crude Accountability. According to the Western Kazakhstan Oblast Ecology Department representative Khon, KPO regularly

monitors four to six airborne toxins. When asked by hearing participants how the toxins were selected, he did not answer.

22. The results of the analysis are available from Crude Accountability by request: P.O. Box 2345, Alexandria, VA 22301.

23. Each of these letters is available on Crude Accountability's Web site: http://www.crudeaccountability.org/.

24. E-mail correspondence from Svetlana Anosova to Kate Watters, July 31, 2006.

25. The author is in ongoing correspondence with the IFC compliance officer.

26. All the documents related to the CAO's investigations at Karachaganak, including the audit, are available on the CAO Web site: http://www.cao-ombudsman.org/html-english/complaintKazkhstanCompliance.htm/ (accessed May 15, 2008).

27. Ministry of Health of the Republic of Kazakhstan [Ministerstvo zdravookhraneniia respubliki Kazakhstan] to Svetlana Anosova and Green Salvation, August 8, 2006 (no. 07-21-6887).

28. Berezovka Initiative Group, personal conversation, May, 2007.

29. According to the editor of *Uralsk Weekly*, the newspaper has been threatened with closure on numerous occasions. KPO has attempted to pay them to publish positive articles about their operations.

References

Agyeman, Julian, Robert D. Bullard, and Bob Evans, eds. 2003. *Just Sustainabilities: Development in an Unequal World*. Cambridge, MA: MIT Press.

Akhmedyarov, Lukpan. 2004. "The Constitution as a Tool of Great Use to Berezovka Residents Fighting for their Rights." *Uralsk Weekly*, December 15. http://www.crudeaccountability.org/ (accessed January 12, 2007).

Akhmedyarov, Lukpan. 2006. "Karachaganak's Sanitary Protection Zone Must Be Expanded." *Uralsk Weekly*, April 6. http://www.crudeaccountability.org/ (accessed January 12, 2007).

Bacheva, Fidanka, Manana Kochladze, and Suzanna Dennis. 2006. "Boomtime Blues: Big Oil's Gender Impacts in Azerbaijan, Georgia and Sakhalin." CEE Bankwatch Network and Gender Action, September.

Belov, Yu., and I. A. Khalii. 2005. "Analysis of the Sociological Research, Resident Opinions: Living Conditions in the Village of Berezovka and Future Scenarios." In *The Residents of the Village of Berezovka, Kazakhstan Consider Resettlement Necessary Given the Harmful State of the Environment, Sociological Survey Research Result*. Institute of Sociology, Russian Academy of Sciences. November. http://www.crudeaccountability.org/ (accessed January 12, 2007).

Bullard, Robert D., ed. 2005. *The Quest for Environmental Justice: Human Rights and the Politics of Pollution*. San Francisco: Sierra Club Books.

Burlinsky Vesti. 2004. "Resettlement: On What Grounds?" December 28. http://www.crudeaccountability.org/ (accessed January 12, 2007).

Center for State Sanitary-Epidemiological Inspectorate in Orenburg Oblast (CSSEIOO) [Tsentr gosudarstvennogo sanitarno-epidemiologicheskogo nadzora v Orenburgskoi Oblasti]. 2004. Report, December 16, Orenburg, Russia.

Chevron. http://www.chevron.com/countries/kazakhstan/ (accessed September 12, 2008).

Compliance Advisor Ombudsman (CAO). 2005. "Assessment Report Complaint Regarding the LUKoil Overseas Project (Karachaganak Oil and Gas Field)." BurlinskyRaion, Western Kazakhstan Oblast, Kazakhstan, April 15.

Compliance Advisor Ombudsman (CAO). 2008. "CAO Audit of IFC, Karachaganak Project, Case of Residents in the Village of Berezovka."

Crate, Susan A. 2006. *Cows, Kin and Globalization: An Ethnology of Sustainability.* Lanham, MD: Alta Mira Press.

Crude Accountability. 2003, December. *Ever Decreasing Circles.* Directed by Mark Warford, Rockboat Pictures, Running Time: 16:26, DVD.

Crude Accountability. 2007. "Turkmenistan's Environmental Risks in the Era of Investment in the Hydrocarbon Sector." September. http://crudeaccountability .org/ (accessed February 29, 2008).

Dashkov, Denis. 2004. "Unbelievable Incident of Great Significance to Berezovka: Local Authorities Refuse to Explain the Reasons for a Police "Attack" on Local Residents." *Respublika*, December 24. http://www.crudeaccountability .org/ (accessed January 4, 2007).

DOJ. See U.S. Department of Justice.

IUCN. See World Conservation Union (IUCN).

Karl, Terry Lynn. 1997. *The Paradox of Plenty: Oil Booms and Petro-States.* University of California Press.

Kenzhegalieva, Gulmira. 2006. "Shefstvo Amerikantsev v Berezovke [Sponsorship of the Americans in Berezovka]." Gazeta.ru, March 13, http://www .sngnews.ru/frame_article/17/58910.html/ (accessed January 12, 2007).

King, Linda Price. 1999. *Chemical Injury and the Courts: A Litigation Guide for Clients and Their Attorneys.* Jefferson, NC: McFarland.

King, Linda Price. 2004. "Popular Epidemiology: Women Lay Scientists Using Participatory Action Research to Solve Environmental Health Problems." B.A. thesis, Union Institute and University, Cincinnati.

Korina, L. 2005. "Kazakhstan: Hydrogen Sulfide and a Conflict of Interest in Berezovka." *Kazakhstanskaya Pravda*, January 29.

Lazko, A. E., and M. V. Lazko. 2005. "Gomestaza u prozhivaushchikh v zone vozdeistviia krupnogo gasoneftepererabatyvaiushchego kompleksa [Homeostasis among Residents in the Impact Zone of a Significant Gas and Oil Refining Complex]." Report, Orenburg, Russia.

Mainhardt-Gibbs, Heike. 2006. "Azerbaijan's Continued Struggle with Poverty and Oil Dependence: Concerns surrounding a Decade of IFI Lending." Bank Information Center Discussion Paper. Washington, DC: August.

Novotny, Patrick. 1998. "Popular Epidemiology and the Struggle for Health in the Environmental Justice Movement." In Daniel Faber, ed., *The Struggle For Ecological Democracy: Environmental Justice Movements in the United States*, 137–158. New York: Guilford Press.

Okonta, Ike, and Oronto Douglas. 2001. *Where Vultures Feast: Shell, Human Rights and Oil in the Niger Delta*. San Francisco: Sierra Club Books.

Ryder Paul, ed. 2006. *Good Neighbor Campaign Handbook: How to Win*. Lincoln, NE: iUniverse.

Sokovnin, Nikolai. 2003. "Dom po imeni Tungush [The House by the Name of Tungush]." *Uralsk Weekly*, March 20.

Tsalik, Svetlana. 2003. *Caspian Oil Windfalls: Who Will Benefit?* New York: Open Society Institute.

U.S. Department of Justice (DOJ). 2007. "Baker Hughes Subsidary Pleads Guilty to Bribing Kazakh Official and Agrees to Pay $11 Million Criminal Fine as Part of Largest Combined Sanction Ever Imposed in FCPA Case." April. http://www.usdoj.gov/opa/pr/2007/April/07_crm_296.html/ (accessed March 9, 2008).

U.S. Securities and Exchange Commission (SEC). 2007. "SEC Charges Baker Hughes with Foreign Bribery and with Violating 2001 Commission Cease-and-Desist Order." http://www.sec.gov/news/press/2007/2007-77.htm/ (accessed March 8, 2008).

World Conservation Union (IUCN). 2007. Sturgeon Specialist Group 1996. *Huso huso* (Caspian Sea stock). In 2007 IUCN Red List of Threatened Species. http://www.iucnredlist.org/ (accessed March 19, 2008).

Zaharchenko, Tatiana, and Gretta Goldenman. 2004. "Accountability in Governance: The Challenge of Implementing the Aarhus Convention in Eastern Europe and Central Asia." In *International Environmental Agreements: Politics, Law and Economics*, 1–24. Dordrecht, Netherlands: Kluwer Academic.

Zlobina, Alla. 2005. "Environmental Dregs." *Uralsk Weekly*, April 7. http://www.crudeaccountability.org/ (accessed January 12, 2007).

8

Viliui Sakha of Subarctic Russia and Their Struggle for Environmental Justice

Susan Crate

This chapter uses an anthropological case study of Viliui Sakha, one of northern Russia's native peoples, to illustrate the environmental injustices prevalent in post-Soviet Russia. Anthropological case study analysis expands the environmental justice agenda by bringing to light the dynamic interplay of culture, power, and the environment and weaving together historical data, sociocultural analyses, and ethnographic voice (Johnston 2001, 132). Brought into an environmental justice frame—integrating the issues of racism, elitism, and economic disparity with unequal exposure to environmental hazards, marginalization in environmental decision-making processes, and lack of voice and remuneration—such case study analysis provides a powerful lens to bring issues of human rights and justice to the fore. It is especially pertinent to the case at hand because Viliui Sakha citizens protesting socioecological injustices do not frame their cause as "environmental justice," despite the leverage that might give them.

After briefly reviewing the history and geopolitical situation of Viliui Sakha, the environmental history of the Viliui regions, and citizen activism and corporate activities there, I contextualize the Viliui Sakha case within some of the relevant research on Russia's other indigenous peoples and also compare it with a case of indigenous peoples and diamond mining in the Northwest Territories (NWT), Canada.[1] I argue that, because the Russian government continues to sell out the environment to promote economic growth, environmental justice can only come about for Russia's inhabitants if there is (1) an informed, empowered, and representative citizenry; (2) an effective legal infrastructure to implement and enforce laws; and (3) a transparent economic development process founded upon comprehensive protocols for environmental, social, and economic sustainable development. Each of these three criteria, in turn, will only be met if civil society continues to expand and the momentum of international precedents and collaboration continues to build.

Background on Viliui Sakha

Between 1400 and 1917, the Russian Empire expanded a thousandfold, from 20,000 square kilometers (7,700 square miles) to 20 million square kilometers (7,700,000 square miles; Shaw 1999). In the process of such extensive "land grabbing," Russia, by default, emerged as an ethnically diverse country with over one hundred peoples represented within its vast borders. Sakha were one of those peoples, Turkic-speaking native agro-pastoralists who adapted their horse- and cattle-breeding subsistence to the subarctic environment of the western Sakha Republic in northeastern Siberia. Sakha's Turkic ancestors migrated from Central Asia to the shores of Lake Baikal in the 900s and then traveled north, from the thirteenth century, following the Lena River to Sakha's present home (see figure 8.1). Viliui Sakha are a Sakha group whose Turkic

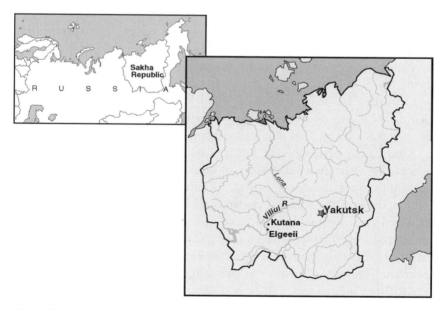

Figure 8.1
On the left, map of the Sakha Republic within Russian Federation; on the right, map of the Viliui Sakha regions with regional center, Suntar, and two research villages, Elgeeii and Kutana.

ancestors settled in the western Viliui watershed regions of the republic. Russians began to colonize these areas in the mid-1600s, annexing Sakha lands and demanding a fur tax; they occupied the native lands for many centuries of the tsarist era and maintained a fairly stable colonial society, from which they launched joint Sakha-Russian trading and missionizing expeditions. Soviet-era collectivization and industrialization radically transformed Viliui Sakha's subsistence practices and natural environment. Today, rural Viliui Sakha depend on a mixed cash economy supplemented with domestic and wild food production; those living in urban centers depend on salaries and wage labor.

The Sakha Republic (called the "Yakut Autonomous Soviet Socialist Republic" or "Yakut ASSR" in the Soviet era) is twice the size of Alaska and rich in mineral wealth and natural resources.[2] Sakha have a folk legend that explains how all the elements of Mendeleyev's periodic table came to be found within their republic's borders. The gods were flying over the earth, the legend goes, giving out all the natural resource wealth from an enormous chest, but when they got to the Sakha regions, it was so cold they froze, and the entire contents of their chest spilled out. These resources, whose exploitation began in full during the Soviet era, today provide the republic and the Russian Federation with sizable revenues (Tichotsky 2000). Russia is the world leader in proven reserves of diamonds, and each year accounts for 20–25 percent of world diamond production; 82 percent of its diamonds are found in the Sakha Republic (Matveev 1998; State Committee for the Environment 2000).

In the first decade of post-Soviet Russia, the Sakha Republic was unique as an emerging economic power, with strong ethnic representation in its governmental apparatus. Unlike other post-Soviet non-Russian peoples, Sakha emerged on an equal, in some cases on a higher than equal, footing with their Russian counterparts in controlling their republic's government and society (Balzer and Vinokurova 1996). One sign of this was Sakha's success at bargaining for a significant percentage of diamond revenues during privatization, shortly after the 1991 breakup of the Soviet Union. The republic was celebrated for its unprecedented environmental record, based largely on former President Nikolaev's granting protected status to 20 percent of the republic's land. Viewed from the West, from Moscow, and from its capital city, Yakutsk, the Sakha Republic appeared to have altogether overcome the hardships of the post-Soviet transition with its robust economy, and its cutting-edge environmental policies.[3]

However, there is a darker side to this story. It involves the environmental and sociocultural impact of the republic's achievements on the rural Sakha inhabitants living close to its main industrial activities, in particular for our purposes, those of diamond mining. Research to date in the Viliui regions addresses the colonization of native populations (Tichotsky 2000), the health impacts of industrial development (Marples 1999; Espiritu 2002), the cultural repercussions of diamond production (Argunova-Low 2004) and the effects of the organizational changes within the ALROSA mining company (Yakovleva and Alabaster 2003; Yakovleva et al. 2000). However, it lacks a comprehensive analysis of the local impacts of diamond mining or the citizen activism that has arisen in response to them in the Viliui regions (Crate 2006). Analyzing the effects of diamond mining in an environmental justice frame can clarify what is needed to expedite improvements in the lives of Viliui Sakha.

The Historical Foundations of Viliui Environmental Issues

Viliui environmental issues arose in the Soviet period from the consolidation of agriculture, first through forced collectivization, then through state farm agro-industrial production and the industrial exploitation of diamonds. For Viliui Sakha, this consolidation dramatically changed indigenous settlement patterns, from extensive to concentrated. Born and raised in areas outside modern villages, contemporary elders remember collectivization involving a series of moves to agro-industrial centers, which left all their formerly productive areas abandoned (Crate 2002). Collectivization also changed Viliui Sakha's lifeways, from indigenous, time-tested, and ecologically based subsistence practices, which made use of vast areas of land and relied on traditional family-clan interdependence, to industrial production, with its dependence on modern transportation to reach necessary resources. In the state farm (sovkhoz), labor was reorganized to realize the sole objective of producing meat and milk for the regional diamond-mining industry, essentially colonizing Viliui Sakha as industrial servants (Crate 2006; Tichotsky 2000). These changes brought with them the loss of indigenous ecological knowledge, the breakdown of the traditional family-clan system, and the environmental stress of living in more populated and polluted areas. On the other hand, it is important to note that there were also many improvements in the lives of Viliui Sakha during the Soviet period,

Figure 8.2
The first diamond pipe, Mirnyi, Sakha Republic, with the city of Mirnyi visible on the horizon.

including literacy, medical care, improved sanitation and hygiene, social services, and access to consumer goods.

The major changes brought about by Soviet-era industrialization in the Viliui regions were a result of diamond mining that began in the late 1950s, shortly after diamonds were first discovered (see figure 8.2). Like all Soviet-era industrialization, the exploitation of diamonds was conducted without environmentally protective measures.

The establishment and operation of the diamond-mining industry gave rise to a variety of Soviet-era environmental offenses in the Viliui regions. First, there was and continues to be contamination of the watershed from phenol, thallium, and highly mineralized brine. To generate sufficient electricity to process the diamonds, the government built the first hydroelectric station on the permafrost. The station's reservoir flooded 2,500 square kilometers (965 square miles) of uncleared taiga, pasture, and swamp, and the resultant anaerobic decomposition of vegetation released a surge of phenols and copper into the watershed. This was further contaminated by a highly toxic thallium-containing compound called "klerich," used to separate diamond granules from their natural substrate, kimberlite. Additionally mineralized brine, high in salts, copper, chromium, nickel, iron, zinc, and lead, seeps up from under the permafrost layer and collects at the base of mined diamond pipes (see figure 8.2). Until 1986, this toxic brine was dumped directly into the Viliui watershed system.

Another Viliui regional environmental offense of the Soviet period was a series of secret underground nuclear tests, performed between 1974 and 1987 by the Soviet government for "civilian purposes" and to build dams for diamond industry holding ponds. Two tests resulted in catastrophic above-ground radioactive fallout. In 1987, a first-ever newspaper article about the blasts generated persistent citizen inquiry. The state had to publicly acknowledge the tests and announce that two, "Kristall" and "Kraton-3," had released catastrophic amounts of nuclear fallout (Burtzev 1993). The lesser of the two, Kristall, occurred in 1974, just 2.5 kilometers (1.5 miles) from the industrial town of Udachnyi. It was the first of eight explosions planned to free the subsoil of permafrost so that a dam could be built for one of several waste filtration ponds of the Udachnyi diamond industry. To accomplish this, the company's engineers sited the detonation points at an unusually shallow depth—99 meters (325 feet). After the catastrophic fallout of the first explosion, the remaining seven were canceled. The second and more destructive of the two, Kraton-3, occurred in 1978, a mere 183 meters (600 feet) from the Markha River, a major tributary of the Viliui and a principal source of drinking water for the villages that border it. Kraton-3 resulted in a massive release of 19 kilotons of radioactive fallout (Burtzev and Kolodoznikova 1997), making it more powerful than the bomb dropped on Hiroshima in 1945 (Pavlov and Afanaseeva 1997).[4] Many specialists in nuclear contamination consider the overall situation in Sakha, with radiation fallout contaminating the soils, air and water as "constituting one of the most serious problems in the history of nuclear explosions" (Yablokov 1992: 11).

In 1990, state specialists and researchers from the Russian Academy of Sciences investigated the situation and found cesium-137, strontium-90, and plutonium-239 and -240 in the soil, rain, lichen, tree bark, and adjacent water systems (Pavlov and Afanaseeva 1997). They verified plutonium contamination at levels equal to the maximum measured for soils in Byelorussia and Ukraine following the 1986 Chernobyl accident (Yablokov 1992). There was no recording of contamination levels when the Kraton-3 explosion occurred in 1978. Local inhabitants were not informed of the extent of plutonium contamination until 1993 (Pavlov and

Afanaseeva 1997). Despite the life-threatening nature of these accidents, there has yet to be a comprehensive study of these two sites.

Other Soviet-era environmental offenses include the three-stage damming of the Viliui River to generate hydropower, which, by altering the temperature and the natural ebb and flow regime of the river, adversely impacted fish populations; the relocation of native settlements from the reservoir flood areas; regular showers of radioactive debris from the Baykonur Cosmodrome second-stage rocket drop-offs (Crate 2003c). Although, for decades, there was no official information about these environmental offenses, local inhabitants witnessed an increase in many forms of human diseases, including cancers, along with a decrease in longevity, a loss of valuable fish species in the river and its tributaries, changes in climate, decreases in wildlife populations, and the death of large areas of forest. One of the main concerns today in the Viliui regions is that of access to clean drinking water. Despite substantial funds for environmental rehabilitation, this issue is yet to be addressed.[5]

The Historical Foundations of the Viliui Environmental Movement

In the late 1980s, with the advent of glasnost and perestroika, inhabitants across the Soviet Union first gained access to information about the environmental offenses of the Soviet period. Empowered by this information, like so many elsewhere in the former Soviet Union, concerned citizens and representatives of Yakutsk's scientific intellectual community founded the Public Ecological Center in early 1989 to research and disseminate information about the environmental legacies of their homeland. In the fall of that year, the center organized a republic-wide conference to discuss ecological problems and to form regional watershed-based organizations to initiate local activism. Many concerned residents of the Viliui regions attended, and there was much discussion of and interest in the complex of environmental issues on the Viliui River. Formed by Nyurba residents concerned about the Viliui watershed, the Viliui Committee, in concert with the Public Ecological Center, organized a conference in the diamond-mining center Mirnyi to meet with representatives of the Almazy Rossii-Sakha Company Ltd., known today only by its acronym, ALROSA (Crate 2006, 2003c).[6] From its participation in the Public Ecological Center in Yakutsk, the committee already had substantial support of many government representatives and had received extensive coverage in the press. Lyubov

Yegorova, one of the original founders, described the committee's work in the following terms:

Our press supported us and played a big role in getting the word out. Soon the whole republic knew about the "Tragedy of the Viliui," and all the other things they called it....We kept talking about the problems. Then in 1991, with the help of the Committee to Save the World and the Sakha Minister of Ecology, we shot a film about the radiation problems and all the ecological problems. We were working. We gathered a lot of money by showing the ecological problems throughout the republic.[7]

During this time, concerned citizens throughout the Viliui regions initiated local chapters of the Viliui Committee and began taking action by organizing village meetings and discussing environmental concerns and contentions. At first, these public meetings were largely soapbox venues—a time to gather the local citizenry and express anger at and disdain for the damage done. As the committee matured, however, regional representatives organized scientific conferences and drafted citizens' petitions to the government. In 1993, Pyotr Martinev, experienced in the technology of diamond transport and passionate about citizen advocacy, joined the committee. Martinev soon became a guiding member, traveling with most of the committee's ecological expeditions and researching the nuclear accident sites in the Viliui regions. He also traveled frequently to Yakutsk to meet with representatives in the republic's parliament. It was with his vision and under his leadership that the committee became actively involved in the legal process.

In 1994, ALROSA announced its discovery of "the biggest diamond column in the Viliui regions," located near several indigenous communities in the Nyurbinskii Raion. While assessing the area, the company discovered a second column. Nearby inhabitants openly voiced their opposition to these new mines, claiming that their settlements already had their share of the environmental havoc wreaked by previous mining activity. Based on the company's record, most did not believe its promise to abide by full environmental protection, including impact statements and monitoring. Although the Ministry for Nature Protection adopted higher environmental standards in the mid-1990s, promising extensive monitoring of the new diamond-mining area to establish a preproduction "before" baseline for contamination levels (a step not taken with the first mine sites forty years earlier), most inhabitants did not believe this promise either.[8] In fact, the ministry would later curtail monitoring due to a lack of funds (Crate 2006; 2003c). And, to this day, it is unclear what tangible changes have been made from previous mining practices, if any at all.

Martinev was a lifelong resident of the Nyurba region and focused all of his efforts toward blocking these new mining ventures. He took every opportunity to argue openly that these new diamond reserves should be mined only after ALROSA had adopted environmentally safe technology and assigned a fair share of the diamond profits to Viliui inhabitants. In 1996, he spearheaded several referenda to these ends and demanding that, based on its failure to perform comprehensive environmental impact assessments, ALROSA be barred from exploiting the new diamond column. But these efforts only temporarily delayed the company's plans.

With Martinev's untimely death in 1997 after several years' struggle with liver cancer, the Viliui Committee suffered a severe setback. Martinev was the guiding vision and force behind the committee. His energy inspired others and his organizational skills cleared and maintained a path for all to follow. After his death and a series of failed referenda, most of the original Viliui Committee members left the organization. During this time, ALROSA launched a propaganda campaign aimed at squelching any remaining environmental objections.

Between 1997 and 1999, Viliui citizens were told outright that, if they advocated too strongly for a clean environment, they would risk losing their state salaries, subsidies, and pensions. From my field research, it was evident that this propaganda campaign worked. I witnessed an active and concerned citizenry turn into a silent and apathetic one over the course of those two years. The regional representatives of the Viliui Committee changed markedly. On the eve of the committee's tenth anniversary in 2000, a complete turnover of membership had produced new priorities that were anything but environmental activism. The new members, all key figures in regional economic development, transformed the Viliui Committee from an environmental NGO focused on involving the citizenry in environmental activism to a bureaucratic board of local officials who gathered privately to discuss their plans. In short, the active environmental agenda of the original Viliui Committee had been successfully co-opted (Crate 2003c).

Contextualizing Viliui Sakha's Plight

Viliui Sakha are not alone in their plight. Across Russia, many indigenous peoples are struggling to resolve the environmental issues of the vast territories upon which they depend. Infringements on indigenous lands for resource extraction and their contamination by oil spills, radioactive leaks, and surface water pollutants continue to threaten the contemporary cultural survival of Russia's indigenous populations (Crate 2003c; Kohler and Wessendorf 2002; Forbes 1999; Wiget and Balalaeva 2000; Wiget and Balalaeva 1997; Metzo, chapter 5, this volume).

Now, at the beginning of the twenty-first century, the situation of Viliui Sakha, like that of Russia's other native peoples, has not gotten better, in part due to continued economic and political instability. As mentioned above, following the 1990s citizen involvement in environmental issues, Viliui Sakha became decidedly apathetic about protecting their local ecology, largely due to governmental threats. With President Medvedev following through with former president Putin's emphasis on natural resource development as Russia's primary source of economic recovery and with much of the country's resource wealth found in indigenous areas, it appears that environmental issues will take a low priority in the nation as a whole (Peterson 2001, 2002).

This is clearly the case for Viliui Sakha. In the last several years, the Russian government has been maneuvering for more control over ALROSA. Its maneuvers have raised widespread citizen protests in the Sakha capital, Yakutsk.[9] It is also rumored, not without some plausible basis, that the Sakha Republic might become a district of lower status, with the objective of increasing the federal share of the republic's resource wealth (see http://www.knia.ru/news/489.html; accessed 9/6/08).[10] A greater federal presence would limit the ability of Sakha not only to reap benefits from their natural resources, but also to control their development. With the autonomy of the entire Sakha Nation at stake, there are clear issues of environmental justice and human rights that need to be brought to the fore.[11]

For Viliui Sakha, the past and current environmental abuses that threaten human health and livelihoods in the Viliui regions, coupled with the loss of sovereignty, constitute a strong case for environmental injustice (Agyeman et al. 2003; Pellow and Brulle 2005; Sandler and Pezzullo 2007). The question becomes to what extent could the fight of Viliui Sakha (and of other post-Soviet indigenous groups struggling with environmental abuses) become stronger by using environmental justice as a frame? The environmental justice model does indeed fit Viliui Sakha's plight if we understand the model to include racism, elitism, and economic disparity as factors in the "unequal siting of environmentally undesirable land uses, routine marginalization from environmental decision-making processes, and denial of just compensation and informed consent in environmental matters" (Sandler and Pezullo 2007, 8).

There are instances that suggest environmental justice is coming into the vocabulary and consciousness of post-Soviet citizens and activist groups. Since the breakup of the Soviet Union in 1991, concerned citizens of Russia have made substantial progress, with the assistance of international NGOs, in building civil society, considered the essential first step to successful environmental movements generally (Soubotin 2002; see also Watters, chapter 7, this volume), and to a successful environmental justice movement—"a local grassroots or 'bottom-up' community reaction to extended [environmental] threats" (Agyeman 2005, 1) —in particular. The key to the environmental progress in Russia is the presence of a strong urban base and of international representation. Even a cursory look at the Web pages of U.S. environmental NGOs such as the Initiative for Social Action and Renewal in Eurasia (ISAR), Pacific Environment, Sacred Earth Network (SEN), and World Wildlife Fund (WWF) reveals how crucial this international-urban connection is.

An exemplary historical case of redressing environmental injustices against an indigenous people is the Russian Association of Indigenous Peoples of the North (RAIPON)'s success in establishing the Tkhsanom Specially Protected Territory of Traditional Nature Use (TTNU), located in the Koryak Autonomous Okrug of Kamchatka in Russia's Far East and encompassing 2.1 million hectares (5.2 million acres, or 8,100 square miles; Murashko and Zaporotsky 2002).[12] Critical to the 1998 establishment of the TTNU were local leadership and vision, the knowledge and use of environmental and indigenous legislation to protect the area, the involvement of the Itelmen Revival Council, the local indigenous RAIPON group and RAIPON's Moscow representation, and support by international organizations, most notably the World Wildlife Fund.[13]

However, in the swiftly shifting tides of Russian internal politics and economic reform, on March 14, 2001, Vladimir A. Loginov, the newly appointed governor of the Koryak Autonomous Okrug, repealed his predessor's executive order establishing the Tkhsanom TTNU, thus opening the area to industrial development and exploitation of its natural resources. This was met with strong protest, and the area's protection was questionable, at best. Central to continued protection of the area are legal coordination between the Tkhsanom communities through their work in the Itelmen Revival Council, RAIPON, and international pressure to make use of the new federal laws that ensure their rights to the historical lands of their ancestors and to protect their environment.

For this case, the necessary components to redressing environmental injustices were a strong urban base, international contact, local leadership, and a knowledge and ability to take advantage of existing legislation. These are the same characteristics that predated the redressing of environmental injustices in other parts of the circumpolar north (Young and Osherenko, 1993). However, given the frequently shifting economic and political tides of contemporary Russia, successful cases like the one cited above are rare.

Insight from Other Circumpolar Countries

The increasing globalization of the world capitalist economy, with production and finance organized on a transnational basis, has been central to the spread of capital-intensive high-value, world market–oriented extractive industries. It has also been central to the growth of informed and sophisticated indigenous movements in affected areas. With the increased focus on sustainable development of both renewable and natural resources in an age when world demand for those resources increases daily, the Arctic presents an especially provocative case. Some arctic indigenous peoples, in defense of their lands and their livelihoods, and acting in concert with scientists, policy makers, and other indigenous peoples, are striving to ensure a workable participatory approach toward achieving sustainable and equitable futures. The case of diamond mining in Canada provides insight into how Viliui Sakha and other indigenous inhabitants of Russia confronting economic dependence and environmental degradation from regional mining could frame their plight as an environmental justice issue.

There are both commonalities and contrasts between the Russian and Canadian cases. Both Viliui Sakha and Canadian Native peoples (and all indigenous inhabitants throughout the circumpolar north, for that matter) are affected by the popular stereotypes of the north, a world area seen as both a resource frontier and a centuries-old homeland for thriving indigenous cultures. Both regions have similar histories of colonization based on a fur trade. Unlike the Russian case, however, the Canadian government has granted each of Canada's Native peoples an official title. Named and written into British common law in the mid-eighteenth century and more fully recognized starting in the mid-twentieth century, the legal developments enforcing land claims and indigenous rights have resulted in indigenous awareness and, in many cases, successful interventions to northern resource development.

Additionally, indigenous activism developed in an entirely different way in Canada, through the Mackenzie Valley Pipeline process and the complex, contradictory evolution of the Assembly of First Nations and the Déné Nation.[14] Community-based resource schemes, comanagement, self-government, and self-determination are found throughout Canada and influence people's behavior and its environmental impact (Berhout et al. 2003, 22). Although both Russia and Canada possess diamonds, their civic and political histories are very different. With these similarities and differences in mind, the following analysis will explore the Canadian case to unearth possible lessons for Viliui Sakha.[15]

The Case of Canadian Diamonds

Charles Fipke discovered Canada's first kimberlite pipes in 1991 under a small lake adjacent to Lac de Gras in the NWT. Fipke had the financial backing of Australian mining giant Broken Hill Propriety (BHP) and formed his own exploration company, Dia Met Mineral Ltd. (Bielawski 2003, 29). This initial discovery of diamonds in arctic Canada in 1991 triggered the NWT "diamond rush," the largest staking rush in North American history. There are presently two working mines, Ekati, which began operation in 1998, and the Diavik diamond mine, which began production in early 2003. Both these operating mines are located approximately 300 kilometers (185 miles) north of Yellowknife, the capital of the NWTs. DeBeers is operating a third mine near Snap Lake.

Because of these mining activities and the resulting employment and service contracts they have brought for northern inhabitants, the NWTs have gone from having the highest unemployment rates in Canada, when the fur trade collapsed in the 1980s under pressure from animal rights groups, to an unemployment rate of 6.1 percent, the lowest in the nation. In the territories' villages, new trucks and snowmobiles fill the driveways, and dozens of Native families are building new homes beside their tipis, where they continue to dry caribou and whitefish. The long-term repercussions of this rapid socioeconomic development should be of concern. Between the 1970s and the mid-1990s, Native inhabitants of Canada's NWTs involved in the lengthy Mackenzie Valley Pipeline hearings on oil and gas development, indigenous rights, and the first land claims of 1984, had transformed themselves from complacent spectators looking on while their land was used for the benefit of others to active, politically empowered stakeholders. When BHP arrived in the NWTs, they were met by Native communities who were both expected and empowered to negotiate directly with them. Native peoples obtained some of the best legal and negotiation advice available.

Before BHP could begin planning the logistics of mining diamonds in the Arctic, it was required to negotiate with the Canadian federal government, the government of the NWTs, and with Native peoples living in the impact area of the mine. The company was required to obtain important licenses, including those dealing with fisheries, water regulation, and land use. It had to contract out an environmental impact statement that, in the end, involved eight volumes of reports weighing 64 pounds. All licenses and approvals were finalized in January 1997; then, after eighteen months of camp construction, production began in October of 1998.

The negotiations were riddled with misunderstandings, ambiguities, and uneven compromises between parties. These arose, in part, because most northern Native inhabitants (who were also most of the local and regional stakeholders in the case), even though they favored the mine for the much-needed employment and prosperity it would bring, also wanted the Environmental Assessment panel to recommend clear and comprehensive terms and conditions both to minimize environmental costs and maximize economic benefits to local communities. These demands were heightened with the testimony of Alex Maun, an indigenous resident of Papua New Guinea, who showed how the Australia-based BHP had caused severe environmental damage at its Oki Tedi mine, had not respected indigenous rights, and had even drafted language that made it a criminal offense for those affected by the mining operation to press charges against BHP (O'Reilly 1996).

In the end, despite their many unresolved deficiencies, by recognizing that social dislocation and other social problems occur when a wage economy is introduced, the impacts and benefit agreements went far beyond any the mining industry had entered into ever before. Both BHP and Dia Met Mineral have been responsive to Native community concerns and have funded programs, changed operating practices, and offered support for advanced education and training.

But what are the costs? Despite the overall economic indicators showing vast improvement in northern Native communities, homelessness, drug and alcohol abuse, and domestic violence remain deep problems among Canada's 1.5 million Native peoples. Financial scandal has plagued many Native communities just as they were gaining broader self-government, and negotiations over old land claims between scores of Native communities and the federal government have bogged down, causing much bitterness. The sudden influx of money has resulted in increased social tensions, financial mismanagement, and higher substance abuse in many Native communities. Although the diamond companies agreed to employ a 44 percent Native workforce, the shift-work patterns that take parents away from children and elders for weeks at a time are not culturally suitable for the social rhythms of Native communities. In turn, because mines have trouble filling their employment quotas of Native workers, they have turned to employing migrant workers, whose growing numbers create even more stress to Native communities.

Issues of employment quotas and migrant populations are further exacerbated by competition between the multinational mining companies involved. There are enough known and probable diamond deposits to keep one or two mines operating for the next hundred years and offering employment for all Native northerners who wish to work in mines. However, DeBeers recently opened a third mine, knowing full well that it cannot possibly recruit enough Native northerners to fill its Native employment quota. If this trend continues, the resources will be mined out quickly, many non-Native southerners will move north, distorting the nature of the Native communities and taxing heavily stressed social systems, and the Canadian North will be challenged again to find ways back to sustainable modes of living.

Environmental impacts, although within the parameters of "anticipated effects" and therefore beyond corporate responsibility, have nonetheless been substantial. Fish habitats have been lost through draining of lakes, destruction of streams, and changes in water quality, with twenty lakes eliminated to date. Land-based habitats for wildlife, including caribou, grizzly bears, and wolverines, have also vanished. There are major differences between the mining companies and indigenous communities in how they understand the environmental impacts of the mines. For example, DeBeers argues that its project at Snap Lake will only change a fraction of caribou habitat. However, caribou do not settle in one place but migrate annually, and areas of mining activity interfere with migratory pastures. Additionally, Native elders warn that caribou will be disturbed by the mines' winter roads, planes, and blasting. Scientists in the NWTs are concerned that the mining companies' vague language, such as "the use of trucks on the haul road will be minimized," will enable them to get away with a great deal more environmental damage than most stakeholders are bargaining for. Finally, the millions of gallons of diesel used to fuel the plants at both mines have produced elevated levels of greenhouse gases. The federal government, apparently short of funds, is many years behind fulfilling its promises to systematically monitor the active mine areas and the areas of exploration.

The more long-term question to pose is just how sustainable is diamond mining for these northern Native communities? Not very, argued a conservation advocate at Environment Canada's summer 2003 workshop on Sustainable Development (Wristen 2003): "Diamond mining, or any sort of mining is clearly not sustainable. You dig a hole, you take stuff out of the hole, and take it somewhere else. Eventually, the hole runs out of stuff you were digging up. That is not sustainable." Thus mining for diamonds, as for other precious minerals, involves the removal and processing of millions of tons of substrate to render small amounts of minerals for a lucrative and highly variable consumer market. Local Native advocates of diamond mining argue that, despite the venture's apparent environmental unsustainability, it contributes to northern Native sustainability more generally by providing economic growth and choices for northern indigenous peoples.

Their view of environmental sustainability may be widely shared among members of the global business community, as reflected in the closing statement of a multinational representative at a DeBeers Snap Lake public hearing: "Functional [biotic] communities will remain in Snap Lake but not to the same extent as is presently the case. The Lake will not be dead but it will be impaired, and this impairment will remain for decades past abandonment, before recovery occurs. And it is likely that the recovery will not result in exactly the same ecosystem as presently exists in Snap Lake. And that, in our view, is the environmental cost of doing business."

There are clear contextual differences between the Canadian and Viliui Sakha cases, such as the presence in Canada of indigenous land claims and their absence in Russia. Over a decade of oil and gas negotiations and more recent interactions with mineral and other interests have made Canadian indigenous groups politically more savvy than their Viliui Sakha counterparts. Finally, because they and their regional and national governments are dealing with multinational as opposed to state-owned companies, they have greater agency to demand environmental protections and economic benefits. Despite these differences, however, the Canadian case informs the striving for environmental justice by Viliui Sakha and other indigenous peoples of Russia to the extent that it sets a precedent that can be translated across international boundaries.

The Case for International Collaboration and Precedent Setting

Researchers have conducted comparative analyses of the environmental and socioeconomic plights of Russia's northern and other circumpolar indigenous peoples for over a decade. Recommendations from such studies emphasize the need to (1) collaborate across disciplines and cultures in joint projects; (2) involve indigenous representatives in the full political processes of local development projects; (3) resolve existing environmental problems through social change; (4) internalize "externalities" to reflect the true cost of development; (5) revolutionize the environmental impact assessment process to encompass holistic approaches; and (6) reorganize the political economy to reflect more egalitarian forms of social development (Chance and Andreeva 1995). One of the major hurdles for Russia's northern indigenous peoples has been the increasing assault on foreign environmental NGOs by the government and by other special interests, which has impeded international collaborations.

Although the case of diamond mining in Canada's NWTs gives rise to further recommendations, it provides no panacea with regard to environmental justice. Indeed, there are widely divergent views on the extent to which Canadian indigenous groups are benefiting from their participation in impact and benefit agreements and the diamond-mining process overall. Some argue that they may be empowered, but that they have used their power to destroy the region's environmental sustainability by promoting mining. Others argue that because stopping the mining altogether was not an option, their winning of concessions was a suitable compromise. For our purposes, the case does provide a clear example of local citizen involvement, indigenous self-determination, and some measure of environmental protection. Local inhabitants were able to organize and demand that companies provide environmental protection and socioeconomic safeguards. In terms of an environmental justice frame,

Déné were engaged in environmental decision-making processes and negotiations for just compensation and informed consent in environmental matters. The Déné's success in these matters was possible because theirs is a civil society where individuals have political know-how within a governmental environment that supports citizen activism and abides by its legal framework. These are key prerequisites for the realization of environmental justice in Russia.

Concluding Remarks

Russia's inhabitants face major impediments to realizing effective environmental justice, even though their environmental plight is the result of racism, elitism, and economic disparity that leave citizens out of decision-making processes, just compensation, and informed consent. Based on this chapter's analysis, the three central criteria cited at its outset to bring environmental justice into being are still lacking in Russia. We can see the closer approximation of environmental justice in the case of diamond mining in Canada's NWTs, which, by meeting those three central criteria, serves as a powerful example for Russia.

The Viliui Sakha study presented here shows that indigenous inhabitants of northern Russia have adapted creatively to the conditions of the post-Soviet transition (Crate 2003a, 2003b; see also Crate and Nuttall 2004). It is the hope of many researchers working with Russia's northern indigenous peoples that these peoples will eventually realize environmental justice, property rights, material compensation, and self-determination at levels witnessed in other parts of the circumpolar north (for example, in Greenland and Canada's Nunavut Territory). Key to this self-determination is involvement with the international community of indigenous groups, with research initiatives, and with governmental bodies elsewhere in the circumpolar north that can facilitate the flow of ideas, experiences, and environmental justice movements across international boundaries. One promising development is the United Nations Declaration on Indigenous Peoples' Rights, adopted by the General Assembly in September 2007—a major international legal instrument that both Russia and the European Union have supported. This and other such international agreements are critical. With continued international collaboration between Russia and its circumpolar neighbors, there is potential for environmental justice to become more and more of a reality for Russia's northern native peoples.

Acknowledgments

I acknowledge all inhabitants of the Viliui regions of the Sakha Republic who have worked with me, and without whose help my research since 1991 would not be possible. I especially wish to thank members of the Viliui Committee, past and present, and all specialists and representatives who gave of their time and expertise concerning environmental issues on the Viliui River. I also gratefully acknowledge my funding sources over the years, in particular the John D. and Catherine T. MacArthur Foundation, the National Science Foundation (NSF), Fulbright-Hays, the Social Science Research Council (SSRC), the International Research and Exchange Board (IREX), and the American Association of University Women (AAUW). Lastly, I thank the book editors, Yelena Ogneva-Himmelberger and Julian Agyeman, the ananymous chapter reviewers, and Jeffrey H Lockridge for their careful work bringing this volume together.

Notes

1. Researchers, policy makers, nongovernmental organizations, and governments often use the term *indigenous* differently (Beteille 1998; Kuper 2003). I use the definition of *indigenous* given in the International Labor Organization's "Convention concerning Indigenous and Tribal Peoples in Independent Countries" (ILO no. 169), "people are regarded indigenous on account of their descent from the populations which inhabited the country, or a geographical region to which the country belongs, at the time of conquest or colonization or the establishment of present State boundaries and who, irrespective of their legal status, retain some or all of their own social, economic, cultural, and political institutions." The Russian understanding of indigenous peoples differs from the global one. It includes twenty-six peoples classified by the Soviets as "small-numbered peoples" in 1925, and another sixteen for a total of forty-two peoples who practice hunting, gathering, and reindeer herding and whose populations do not exceed 50,000 (Slezkine 1994, 2). But it does not include "large-numbered" peoples such as the Komi, Yakut (Sakha), and Buriat (Shnirelman 1999, 119), although these same peoples are classified as "indigenous" in global terms. I use the terms *indigenous* and *native* (or *Native* for North American peoples) interchangeably (see Brown 2003, xiii).

2. The Soviet government designated autonomous regions (oblasti, *okrugi*, and, in some case, republics) for numerically large peoples of the Soviet Union (see note 1).

3. Beginning in 2000, with the end of Sakha Republic President Nikolaev's second 4-year term, his replacement by President Shtrop, and on the eve of Russian

President Putin's first 4-year term, the dynamics of Sakha's ethnic engagement in economic development and mineral exploitation began to change. Since that time, Moscow has gained more and more control of natural resource and mineral exploitation in the republic.

4. Estimated yield of Hiroshima bomb was 15 kilotons, according to the *Encyclopaedia Britannica*.

5. Despite its activities and investments, ALROSA is in many ways "missing the target." Real rehabilitation of the environment will only occur when the environmental damage of the Soviet and post-Soviet periods is not only recognized but also accounted for and remediated. One pervasive issue that illustrates how the many targets of environmental rehabilitation in the Viliui regions are missed, is that of safe drinking water. In the context of a three-year community sustainability project, during focus group sessions discussing the obstacles to future village-level sustainability, local inhabitants were, first and foremost, concerned about the lack of safe drinking water (Crate 2006). They offered countless testimonies about the serious health effects a lack of safe water has meant for them, their communities, and their animals. Although Vasili Alekseev, former minister of ecology for the Sakha Republic, and his regional representatives all agree that the first and foremost need of Viliui inhabitants is good drinking water, most of the funds for environmental rehabilitation of the Viliui River are allocated to develop gas, oil, and high-voltage electricity development there (Vasili G. Alekseev, interview, Yakutsk, Sakha Republic, June 2003; Crate 2006, 2003c). Because of the climate and permafrost, water is in short supply year-round in the Vilniui regions; industrial and agricultural contamination of the watershed has made what little water there is unsuitable for drinking. The diamond-mining industry needs to fund water purification facilities in Viliui watershed rural settlements, as it has done in the urban areas of Lensk, Mirny, Aikhal, Udachny, and Chernyshevsky. Asked why they were not more vocal about having their government provide for safe drinking water and ensure their other basic human rights, most Viliui citizens feared for their salaries, pensions, and subsidies (from my field data for the Nyurbinskii and Suntarskii Raiony, Sakha Republic, 1996–1997).

6. Pyotr N. Martinev, interview, Nyurba, Sakha Republic, July 1997.

7. Lyubov Yegorova, interview, Yakutsk, Sakha Republic, August 2003. Former Sakha president Mikhail Nikolaev also played an advocacy role by writing an article in the Russian newspaper *Moskovskiye Novosti* (July 5, 1992) about the atomic explosions in the Viliui regions.

8. Pyotr N. Martinev, interview, Nyurba, Sakha Republic, August 1996.

9. There are several Web sites that feature these citizen protests. See, for example, http://diaspora.sakhaopenworld.org/alrosa6.shtml and http://www.regnum.ru/news/519845.html/ (both accessed 9/6/08).

10. Ivan Shamaev, interview, Yakutsk, Sakha Republic, August 2005.

11. The 1990 Declaration of Sovereignty of the Republic of Sakha established the republic as a sovereign state within the Russian Soviet Federative Socialist Republic and stipulated that "land, its minerals, water, forests, flora and fauna,

other natural resources, air space and the continental shelf on the territory of the republic shall be its exclusive property." As one of the most activist regions in Russia, the Republic of Sakha was able to establish a special relationship with the federal center and to secure economic concessions, such as a significant share of revenues from the regional production of diamonds and gold. Economic demands and the ability to retain and manage the republic's wealth to alleviate its social and economic crisis lie at the heart of the republic's political agenda (Jackson and Lynn 2002). For Viliui Sakha, three key points should be emphasized. First, although they are able to politically participate in the government of their region, their lack of special status as an indigenous people within the Russian Federation limits their ability to obtain greater concessions. According to legislation, the Russian state provides support and protection to ethnic minorities who consider themselves independent ethnic entities; occupy the territories of traditional settlement of their ancestors; maintain traditional lifeways, economies, and trades; and have a population no greater than 50,000. Because of their large numbers, Viliui Sakha are not recognized as "indigenous." Second, that their attempts to secure regional sovereignty and to build regional power by capturing or sharing in the control over economic resources were not entirely successful further limits their ability to obtain greater concessions. The battle for sovereignty of the Republic of Sakha was dedicated to the acquisition of economic powers that would lift the republic from the status of a resource colony. And third, the special center-periphery relationship with, and the initial liberties given to, Viliui Sakha in the development of their regions came under scrutiny by President Putin. The Constitution of the Republic of Sakha (Yakutia), adopted on April 27, 1992, did not coincide fully with the Constitution of the Russian Federation, adopted on December 12, 1993. Article 72 of the Constitution of the Russian Federation stipulates that the following issues are under joint jurisdiction of the Russian Federation and its subjects (i.e., republics, *kraia*, oblasti, *okrugi*): issues of the possession, use and management of the land, mineral resources, water and other natural resources; delimitation of state property; management of natural resources, protection of the environment and ecological safety; specially protected nature reserves; protection of historical and cultural monuments. President Putin later unified the principles of regional statutes within the Russian Constitution, the main legal framework of the Russian Federation, and, in doing so, the Constitution of the Republic of Sakha was similarly altered.

12. The Russian Association of Indigenous Peoples of the North (RAIPON), founded in 1990, is an association of Russia's "small-numbered" peoples. The accepted limit of "small-numbered," as decided in 1992, is 50,000. There are forty-two such peoples recognized within Russia, whose members total 300,000, with the largest being the Nenets at 35,000 (Kohler and Wessendorf 2002). RAIPON has a central office in Moscow and seeks ways to work with the Russian government. Much of the organization's success results from its international status as a permanent member of the Arctic Council and its special consultative status in the UN Economic and Social Council (ECOSOC). Similarly, it has improved its political and executive structure primarily through funding from

international projects (Kohler and Wessendorf 2002, 26). RAIPON's strategy of action in Moscow and other population centers across Russia includes getting indigenous representatives into positions of public office and influencing state authorities through the dissemination of information (lobbying) and court appeals. In the regions RAIPON focuses on, local representatives facilitate seminars to educate residents on environmental and legal issues, organize negotiations between RAIPON and managers of environmental organizations and industry, and assist in the organization of civil actions and collective appeals in cases of illegal industrial acts. RAIPON also disseminates information about its activities and court precedents and about the infringement and protection of indigenous people's rights through its publication *Indigenous World—Living Arctic*. The full text of the statute establishing the Tkhsanom protected area (TTNU) is available at http://www.faculty.uaf.edu/ffdck/OOPTTP.pdf (accessed 9/6/08)

13. The Itelmen Revival Council "Tkhsanom" (Dawn) has been in existence since February 5, 1989. Based in Kovran, the council officially has 490 members from the Kamchatka Itelmen communities of Kovran, Tigil, Khairiuzovo, Palana, Ossora, Razdolny, Milkovo, and Petropavlovsk-Kamtchatsky. http://www.indigenous.ru/fotki/bull_eng/e_7.htm#itelmen (accessed 9/6/08).

14. "Déné" is the collective name for a widespread group of Athabaskan tribes.

15. I should make a disclaimer here. My analysis of the Canadian case relies on secondary information from key texts and interviews with knowledgeable parties rather than any firsthand fieldwork among the Déné.

References

Agyeman, Julian. 2005. *Sustainable Communities and the Challenge of Environmental Justice*. New York: New York University Press.

Agyeman, Julian, Robert Bullard, and Bob Evans, eds. 2003. *Just Sustainabilities: Development in an Unequal World*. Cambridge, MA: MIT Press.

Argunova-Low, Tatiana. 2004. "Diamonds: Contested Symbol in the Republic of Sakha (Yakutia)." In Erich Kasten, ed., *Properties of Culture—Culture as Property: Pathways to Reform in Post-Soviet Siberia*, 257–265. Berlin: Dietrich Reimer.

Balzer, M., and U. Vinokurova. 1996. "Nationalism, Interethnic Relations and Federalism: The Case of the Sakha Republic." *Europe-Asia Studies* 48 (1): 101–120.

Berhout, F., Melissa Leach, and Ian Scoones. 2003. *Negotiating Environmental Change: New Research in the Social Sciences*. Cheltenham, UK: M. A. Edward Elgar.

Beteille, Andre. 1998. "The Idea of Indigenous People." *Current Anthropology*, 39 (2): 187–192.

Bielawski, Ellen. 2003. *Rogue Diamonds: The Rush for Northern Riches on Dene Land*. Vancouver, BC: Douglas and McIntyre.

Brown, Michael. 2003. *Who Owns Native Culture?* Cambridge, MA: Harvard University Press.

Burtzev, I. S. 1993. *Iadernoe zagriaznenie Respubliki Sakha: Problema iadernoi bezopasnosti* [Nuclear Contamination of the Sakha Republic: The Problem of Nuclear Safety]. Yakutsk: Polygraph.

Burtzev, I. S., and E. N. Kolodoznikova. 1997 *Sovremennaia radiatsionnaia obstanovka na ob"ektakh avariinykh podzemnykh iadernykh vzryvov* [The Contemporary Radiation Conditions at the Sites of Catastrophic Underground Atomic Explosions]. In N. P. Pavlov and V. M. Afanaseeva, eds., *Bol' i tragediia sedovo Viliuia* [The Suffering and Tragedy of the Gray (Ancient) Viliui], 38–41. Yakutsk: SAPI-Torg-Knigi.

Chance, N. A., and E. N. Andreeva. 1995. "Sustainability, Equity, and Natural Resource Development in Northwest Siberia and Arctic Alaska." Human Ecology 23 (2): 217–240.

Crate, Susan. 2002. "Viliui Sakha Oral History: The Key to Contemporary Household Survival." *Arctic Anthropology* 39 (1): 134–154.

Crate, Susan. 2003a. "Viliui Sakha Post-Soviet Adaptation: A Subarctic Test of Netting's Smallholder Theory." *Human Ecology* 31 (4).

Crate, Susan. 2003b. "The Great Divide: Contested Issues of Post-Soviet Viliui Sakha Land Use." *Europe-Asia Studies*, 55 (6): 869–888.

Crate, Susan. 2003c. "Co-option in Siberia: The Case of Diamonds and the Vilyuy Sakha." *Polar Geography* 26 (4): 289–307.

Crate, Susan. 2006. *Cows, Kin and Globalization: An Ethnography of Sustainability*. Walnut Creek, CA: Alta Mira Press.

Crate, Susan, and Mark Nuttall. 2004. "Russia in the Circumpolar North." *Polar Geography* 27 (2): 85–96.

Espiritu, Aileen A. 2002. "The Local Perspective: Interviews with Sakha in the Viliui River Region." *Central Eurasian Studies Review* 1 (1): 15–17.

Forbes, Bruce. 1999. "The End of the Earth: Threats to the Yamal Region's Cultural and Biological Diversity." In *Wild Earth*, http://home.planet.nl/~innusupp/english/forbes2.html (accessed 9/6/08).

Jackson, L., and N. Lynn. 2002. "Constructing Federal Democracy in Russia: Debates over Structures of Power in the Regions." *Regional and Federal Studies* 12: 91–125.

Johnston, Barbara Rose. 2001. "Anthropology and Environmental Justice: Analysts, Advocates, Mediators, and Troublemakers". In Carol Crumley, ed., *New Directions in Anthropology and the Environment*, 132–149. Walnut Creek, CA: Alta Mira Press.

Kohler, Thomas, and Kathrin Wessendorf, eds. 2002. *Towards a New Millennium: Ten Years of the Indigenous Movement in Russia*. Copenhagen: IWGIA.

Kuper, Adam. 2003. "The Return of the Native." *Current Anthropology* 44 (3): 389–402.

Marples, David R. 1999. "Environmental and Health Problems in the Sakha Republic." *Eurasian Geography and Economics* 40 (1): 62–77.

Matveev, A. 1998. "Almazy Rossii-Sakha: Sostoyanie, perspektivy, problemy kompanii [Almazy Rossii-Sakha: Situation, Perspectives, and Problems of the Company]." *Problemy Teorii i Praktiki Upravleniia* [Problems of Theories and Practice of Management] 3.

Murashko, Olga, and Oleg Zaporotsky. 2002. "How the Constitutional Right to Protect the Traditional Environment of Inhabitancy and a Traditional Way of Life can be Implemented." In Thomas Kohler and Kathrin Wessendorf, eds., *Towards a New Millennium: Ten Years of the Indigenous Movement in Russia*, 224–245. Copenhagen: IWGIA.

O'Reilly, Kevin. 1996. "Diamond Mining and the Demise of Environmental Assessment in the North." *Northern Perspectives* 24 (1). http://www.carc.org/pubs/v24no1-4/mining.htm (accessed 9/6/08).

Pavlov, N. P., and V. M. Afanaseeva, eds. 1997. *Bol' i tragediia sedovo Viliuia* [The Suffering and Tragedy of the Grey (Ancient) Viliui]. Yakutsk: SAPI-Torg-Knigi.

Pellow, David, and Robert Brulle, eds. 2005. *Power, Justice and the Environment: A Critical Appraisal of the Environmental Justice Movement*. Cambridge, MA: MIT Press.

Peterson, D. J. 2001. "The Reorganization of Russia's Environmental Bureaucracy: Implications and Prospects." *Post-Soviet Geography and Economics* 42 (1): 65–76.

Peterson, D. J. 2002. "Russia's Industrial Infrastructure: A Risk Assessment." *Post-Soviet Geography and Economics* 43 (1): 13–25.

Sandler, Ronald, and Phaedra Pezzullo. 2007. *Environmental Justice and Environmentalism: The Social Justice Challenge*. Cambridge, MA: MIT Press.

Shaw, D. J. B. 1999. *Russia in the Modern World: A New Geography*. Oxford: Blackwell.

Shnirelman, V. 1999. "Introduction: North Eurasia." In R. Blee and R. Dalys, eds., *The Cambridge Encyclopedia of Hunters and Gatherers*, 119–173. Cambridge: Cambridge University Press.

Slezkine, Yuri. 1994. *Arctic Mirrors*. Ithaca, NY: Cornell University Press.

Soubotin, Xenia. 2002. "Building Civil Society through Environmental Advocacy." Paper presented at Russia's Environment: Problems and Prospects Conference, October 24–26, Miami University, Oxford, Ohio.

State Committee for the Environment [Goskomekologia] (2000). *Gosudarstvennyi doklad o sostoyanii prirodnoi sredy v Rossiiskoi Federatsii v 1999 godu* [State Report on the Environmental Situation in the Russian Federation in 1999]. Moscow.

Tichotsky, John. 2000. *Russia's Diamond Colony: The Republic of Sakha*. Amsterdam: Harwood Academic.

Wiget, Andrew, and Olga Balalaeva. 1997. "Black Snow: Oil and the Eastern Khanty." http://www.nmsu.edu/~english/hc/IMPACTOIL.html (accessed 9/6/08).

Wiget, Andrew, and Olga Balalaeva. 2000. "Alternative to Genocide: The Yuganskii Khanty Biosphere Preserve." http://www.nmsu.edu/~english/hc/ hcbiosphere.html (accessed 9/6/08).

Wristen, Karen. 2003. Adaptation of presentation to Environment Canada's Workshop for Sustainable Development. http://www.carc.org/mining_sustain/ diamonds_arent.php (accessed 9/6/08).

Yablokov, A. V. 1992. "The Recent Environmental Situation in Russia." *Environmental Policy Review* 6 (2): 1–12.

Yakovleva, Natalia, and Tony Alabaster. 2003. "Tri-Sector Partnership for Community Development in Mining: A Case Study of the SAPI Foundation and Target Fund in the Republic of Sakha (Yakutia)." *Resources Policy* 29: 83–98.

Yakovleva, Natalia, Tony Alabaster, and Palmira G. Petrova. 2000. "Natural Resource Use in the Russian North: A Case Study of Diamond Mining in the Republic of Sakha." *Environmental Management and Health* 11 (4): 318–336.

Young, Oran, and Gail Osherenko. 1993. *Polar Politics: Creating International Environmental Regimes*. Ithaca, NY: Cornell University Press.

9

Environmental Justice and Sustainability in Post-Soviet Estonia

Maaris Raudsepp, Mati Heidmets, and Jüri Kruusvall

Environmental justice has two principal aspects: how public environmental resources and environmental benefits and costs are distributed among different groups in the society (its distributive aspect); and how environmental decisions are made in terms of public access and participation (its procedural aspect). In this chapter, as we describe the most notable changes in the social and natural environments of post-Soviet Estonia, we will examine the distributive and procedural aspects of environmental justice as it is perceived there.

Socioeconomic Context: Rapid and Costly Transformation

In 1991, Estonia gained its independence from the collapsing Soviet Union; in 1992, a national currency was introduced. In 1994, the Soviet Army left the country, and a decade later, in the spring of 2004, Estonia joined NATO and the European Union, completing its transition to democracy and a market economy.

Like that of other post-Soviet countries, Estonia's transition to democracy and a market economy was accompanied by economic inequalities, poverty, and social exclusion. Unemployment climbed steeply until 1996, when it leveled off at around 10 percent. It increased again under the impact of the Russian economic crisis in 2000, reaching 13.6 percent. In the last few years, however, the Estonian economy has been extremely successful: GDP increased by roughly 10 percent and the unemployment rate decreased to 5–6 percent in 2005 and 2006. According to the UNDP Human Development Index, Estonia is ranked 40th among the countries of the world—the most developed of the former Soviet republics, but still among the least developed of the EU countries (Heidmets 2007).

At the same time, Estonia has paid a high social price for the success of its radical liberal economic reforms. Since 1991, the birthrate has decreased, mortality increased, and the population declined overall. The number of live births per year fell from a high in 1987and 1988—the beginning of perestroika—of 25,086 and 25,060, respectively, to a current low of about 14,000 live births a year. The average life expectancy at birth, which rose to 71 years in the second half of the 1980s, dropped sharply in the beginning of the 1990s, to 66.9 years in 1994. Particularly noticeable was the drop in life expectancy among men. After declining still further during the mid-1990s, average life expectancy has been gradually rising but still remains the lowest among EU states (Kiivet and Harro 2002).[1] In other eastern European countries, especially Latvia and Lithuania, changes in average life expectancy have been similar to those in Estonia, and are explained by the socioeconomic difficulties related to the transition period.

Rapid reforms in the economy have brought social inequalities and uneven development of regions. The Gini index provides a measure of income or resource inequality within a population.[2] Estonia's Gini index had varied from 0.35 to 0.38 since 1994, remaining among the highest among central and eastern European EU (candidate) countries, indicating huge income disparities (Saar and Lõhmus 2003).

Regional inequalities have only grown worse since 1991. For example, in the year 2000, there was a twofold difference in the average wage between Tallinn and Põlva in southeast Estonia. Rural areas are depopulating and the income of farmers is comparatively low. Unemployment rates also differ notably between various counties, with the highest unemployment rate over three times that of the lowest.

Social reforms were aimed at the abolition of privileges guaranteed by the previous system to certain social groups, most notably, pensioners, previous workers of major industries, and public servants. As a result, many in these groups felt themselves thrown into unexpected insecurity and deprived of their earned position in society.

Estonian society is multicultural. Non-Estonians (over 80 percent of whom are Russian) constitute 32 percent of the population; of these, only half were born in Estonia. Since Estonia regained its independence in 1991, Estonians have become the power-holding majority, whereas a significant number (over 30 percent) of non-Estonians still lack Estonian citizenship and 100,000 hold Russian citizenship as residents in Estonia.

Estonians and local Russians are exposed to significantly different living conditions. For instance, most non-Estonians live in the cities and have less day-to-day contact with nature. Furthermore, those who live in Ida-Virumaa in industrial northeastern Estonia (most of whom are Russian) are more exposed to environmental degradation, especially air pollution, and to difficult work conditions in mining or similar industries than Estonians living elsewhere. Here they continue to live in territorially and culturally concentrated communities and have little contact with Estonians. According to a 2005 social survey (Kreitzberg et al. 2006), Ida-Virumaa is characterized by the worst housing conditions, the lowest household incomes, the fewest cars and computers, and lowest level of access to the Internet.

Post-Soviet Transformations of the Environment

The Estonian natural environment is unique in several ways, with a high level of biological and landscape diversity and with many rare species of plants and animals. More than 60 percent of Estonia's territory is covered by forest and wetlands (Estonia is among the five most forested countries in the European Union). Compared to other industrialized countries, it has a relatively well-preserved natural environment. Furthermore, Estonia has a long history of nature preservation: its first protected area was established in 1910; today, some 15 percent of its territory is in protected areas and nature reserves. Environmental legislation and institutions have successfully advanced the cause of nature preservation.

During the country's post-Soviet economic recession, human environmental impact has diminished. Nevertheless, Estonia has a disproportionately large ecological footprint. Although the population density (31 persons per square kilometer, or 80 persons per square mile) and the standard of living (estimated 2006 GDP per capita: $17,000) are relatively low in the EU context, the country's ecological footprint is more than twice the optimum cap because of its wasteful energy production, based almost exclusively upon oil shale.[3]

Ida-Virumaa in the northeast remains an area of critical concern. Inherited from the Soviet Union, oil shale–burning power plants there are the primary source of Estonia's major airborne pollutant, sulfur dioxide. Due to outdated technologies and individual waste, energy usage is over five times more intensive in Estonia (about 17.8) than the EU average (about 3.4). Despite a major decrease in carbon dioxide emissions in recent years, Estonia still has one of the largest ecological footprints in Europe—more than twice the size of Latvia's or Lithuania's (Merisaar 2007).

Contrary to what its ecological footprint implies, the people of Estonia only consume an average level of goods and services. At the same time, thanks to old, wasteful technologies in local industry, substantial losses in heating and electrical energy transmission systems, and partially subsidized communal services, the consumption of material and energy per unit for these goods or services is much higher than in western European countries. Estonia's wasteful use of resources and energy is the source of its biggest environmental problem—the high level of air pollution due to oil shale power plants and increasing number of cars. As a result, emissions of carbon dioxide per inhabitant are among the highest in Europe. Nevertheless, the quality of the physical environment has become healthier since 1991—air and water pollution have decreased overall, mainly due to a decline in heavy industry.

One-third of the Estonian population lives in Tallinn and its suburbs. The quality of the urban environment depends to a large extent on the quality of urban air. During the last few years, the number of people using cars has grown tremendously. For example, in 1990, 10 percent of the public in Tallinn traveled by car and 90 percent by public transport; by 1997, 60 percent traveled by car and 40 percent by public transport. Similar changes in western Europe took forty years to occur (Merisaar 2007).

Starting in 1991, land reform has led to the privatization of land and forests, with radical changes in the legal regulation of forest ownership. By 2004, 38 percent of the forests in Estonia belonged to private individuals, over 60 percent of whom were original owners whose holdings, once expropriated by the state, had been restored to them (Meikar 2005). According to indirect estimations, there are now 55,000 private forest owners in Estonia, most of whom are urban residents. A peculiarity of Soviet-era land use was the highly centralized kolkhoz system, under which land was intensively used in some regions, but abandoned in others. With land reform, large production units have been broken up into smaller agricultural companies, households, and farms.

A peculiarity of the post-Soviet system, on the other hand, was the recategorization of public resources, under which access to the outdoors went from inclusive to exclusive. Formerly, the environment consisted mostly of public space; collective ownership of common resources dominated, in accordance with the principle that "everything belonged to everybody and accordingly to nobody." Common public space, though socially indeterminate and without a clear owner, was freely accessible to all citizens, who were expected to care and be responsible for the common resources. A major drawback of this system is obvious: private space was reduced to the home environment only. Post-Soviet transformation of the public space meant privatization of formerly common resources, restoration of private ownership, and the establishment of restricted (exclusive) access. As a result, new kinds of conflicts emerged. People used to treating the natural environment as a common good experienced confusion and stress when their access to certain places, coastlines, and roads was limited. Conflicts arose, accompanied by mutual distrust, between the landowners and the general population, primarily over the rights of public access. Although Estonia established the right of free access to natural environments in the 1990s (Ranniku 1996), according to our survey in 2006 (Raudsepp 2007), 50 percent of those asked did not know about it.

Although the environmental impact of the Soviet Army was harmful, the worst degradation was confined to relatively small areas. About one-tenth of Estonian territory, consisting mostly of coastal areas, was a restricted military zone. Despite Soviet military pollution in this zone, plant and animal species remained otherwise undisturbed by human activity for decades.

Perceived Injustice

The main causes of perceived social injustice are related to new market relationships. Employment in privatized enterprises differs fundamentally from the secure lifelong jobs provided by the Soviet state in its industries. For instance, under privatization, structural unemployment has occurred in previously overindustrialized regions, as it did in rural regions after the dissolution of the Soviet kolkhozy and sovkhozy. Regions with high unemployment lost the social capital needed for development, creating a vicious circle that led to the dispossession and social uprooting of people in entire areas, and to the consequent spread of crime, alcoholism, drugs, and HIV.

Since 1991, the pendulum of public opinion on Estonia's liberal reforms has swung from uncritical acceptance to broadening dissatisfaction, with social scientists expressing ever greater concern. This change in opinion reflects the widening disparities in living standards, growing social exclusion, rising unemployment, and a declining population (Lauristin 2003). A survey taken in 2003 (Lauristin 2004) clearly correlated Estonians' general evaluation of their country's transition with its perceived performance on the international level. In contrast with the positive evaluations of the early 1990s, most Estonians expressed dissatisfaction with the outcomes of reforms to improve social justice, employment opportunities, and living standards. A more recent study (Hallik et al. 2006) found that members of ethnic minorities in Tallinn continued to perceive unjust distribution of resources and ethnic discrimination (Hallik et al. 2006).

There have been no special studies on perceived environmental justice as such. Yet some indirect data can be found. In 2002 and 2006, two nationwide surveys were carried out by the Environmental Psychology Research Unit, which focused on various aspects of environmental awareness (Raudsepp 2005, 2007). The samples ($N = 1,000$) were representative of the age, gender, and territorial distribution of the population of Estonia. Self-administered questionnaires in Estonian and Russian were used. Multi-item measures of environmental consciousness were grouped into three blocks: attitudes toward and beliefs about (1) the natural environment (e.g., emotional involvement with nature, perceived restorative qualities of forests, intrinsic value of nature); (2) environmental protection issues (e.g., concern for preserving the environment); and (3) environmentally responsible activities (e.g., perceived social norms, knowledge of environmental consequences, self-reported habitual behavior). In 2006, several questions concerning attitudes toward and beliefs about the forest environment, forest ownership, and new forms of land use were added. The quantitative data were analyzed using an SPSS 10.0 statistical software package.

In the surveys, respondents could specify their environmental concerns. They expressed concern for concrete objects of nature (such as the forest and trees) as well as for the state of the entire ecosystem, based both on direct contacts with nature and on indirect environmental information. Most of their responses mentioned polluted air and water (44 percent of responses), garbage in nature (28 percent), and the destruction of Estonian forests (27 percent). Harming animals and birds (destruction of their habitats) was noted in 8 percent of responses, the deplorable state of nature in the towns and cities (e.g., lack of greenery) was men-

Table 9.1
Personal assessment of the environment by region, Estonia, 2002 and 2006

| | How do you assess the state of the environment at your home locality? | | | |
| | Bad or very bad (%) | | Good or very good (%) | |
Region of residence	2002	2006	2002	2006
Tallinn	38.6	27.1	40.0	25.1
Northwest	26.8	7.8	52.9	37.4
West and islands	10.8	10.9	65.8	58.4
Tartu	14.8	5.6	58.4	46.4
South	16.7	7.7	67.6	53.0
Northeast (Ida-Virumaa)	48.0	21.1	31.1	28.7

Both surveys were conducted nationwide and both used representative samples of 1,000 persons.

tioned in 2 percent of responses. Some respondents (1 percent) also mentioned characteristic features of contemporary Estonia—neglected fields and lands overgrown with brush. Among green issues, forests were the most frequent object of concern—both the irresponsible destruction of forests and their pollution with garbage were perceived as serious problems. The data from several survey questions indicate that the environmental conditions in Ida-Virumaa and in Tallinn are consistently perceived as less favorable than in other regions of Estonia. (See tables 9.1 and 9.2.)

Attitudes toward Participation in Decision Making
According to survey results (Kruusvall 2002), 38 percent of Estonians and 42 percent of non-Estonians do not want to participate in voluntary organizations, while 53 percent of Estonians and 55 percent of non-Estonians do not want to take part in collective protests or petitions. Moreover, 65 percent of Estonians and 67 percent of non-Estonians do not want to be members of any political party.

On the abstract level, non-Estonians are somewhat more willing to be socially active: 41 percent, compared to 36 percent of Estonians, agree with the statement "It is important to participate in public life and make one's contribution." At the same time, there are relatively few

Table 9.2

Personal assessment of recent changes in the environment by region, Estonia, 2002 and 2006

	How has the state of the environment recently changed at your home locality?			
	Deteriorated somewhat or considerably (%)		Improved somewhat or considerably (%)	
Region of residence	2002	2006	2002	2006
Tallinn	35.9	29.6	26.6	38.1
Northwest	27.5	24.3	34.8	30.4
West and islands	18.0	13.0	36.0	54.7
Tartu	23.2	16.3	26.8	41.8
South	28.1	13.9	38.6	45.2
Northeast (Ida-Virumaa)	37.8	25.7	26.7	41.2

Both surveys were conducted nationwide and both used representative samples of 1,000 persons.

non-Estonians in representative bodies. Thus, even though non-Estonians constitute approximately one-third of the population, only 9 percent of local council members and only 6 percent of members of Parliament are non-Estonian (Hallik 2002). Ethnic Estonians are reluctant to allow the equal political participation of non-Estonians: nearly half of Estonians (44 percent) in northeast Estonia would not tolerate more than 10 percent of non-Estonians in the local councils. In Tallinn, non-Estonians constitute 62 percent of apartment tenants but only 54 percent of the members of the tenants' cooperative guiding bodies.

Many non-Estonians, including those who are Estonian citizens, are not able to fully realize their right to participate in the formation and implementation of environmental policy as members of elected bodies. Even though, inspired by idealistic and humanistic values, Estonians and non-Estonians alike took active part in political movements for environmental and social justice in the 1980s, nowadays, more utilitarian and individualistic tendencies prevail (as evidenced by the strong support of right-wing parties among Estonians).

Because the non-Estonian community has not consolidated along ethnic lines (as evidenced by the disappearance of any influential Russian parties), the younger generation needs other avenues for collective participation (beside individualistic forms of self-realization). Taking part in environmental movements could be one such avenue. Although non-Estonians will remain for some time proportionately underrepresented in legislative bodies for various reasons (including a poor knowledge of the Estonian language, lack of Estonian citizenship, and negative attitudes about Estonians), it would be useful for them to develop nonpolitical voluntary associations and NGOs in order to promote their interests and realize their right to participate in environmental decision making.

In sum, according to opinion polls and media analysis, injustice is perceived in relation to human rights, property rights (especially injustice as perceived by apartment tenants on the part of the restored owners of their buildings), social inequality, and similar issues—but *not* in relation to the inequality of environmental conditions.

Environmental Justice: Public Participation

The scope and focus of environmental activism have changed over the last fifteen years. One of the characteristic features of the democratic revival in Estonia in 1990s was the emergence (or reemergence) of various citizen initiatives, social movements, voluntary associations, and groups, all of which sought to attain a particular goal or defend a particular common interest by joint action. Being key structures of civil society, they formed an organizational basis for citizen participation. Thus the first manifestations of participatory democracy (as complementary to representative democracy) emerged in different spheres of life. On the individual and group levels, this meant transition from reliance on the state authorities to self-reliance and mutual help (e.g., local barter exchange, mobilizing community resources). Psychologically, it signified the adoption of specific attitudes and modes of interaction. For example, from the standpoint of an individual, "participation" is not limited to taking initiative and responsibility but also includes the will and ability to engage in joint activity, to tolerate differences and compromises, to sacrifice some individual freedoms for the sake of common interest, and so on. On the other hand, such initiatives tend to function as collective subjects, as localized or network communities.

In 1993, there were approximately 3,500 registered societies and foundations in Estonia. Their relatively great number was partly determined by a long tradition of voluntary organizations and their great cultural significance in the past (Aarelaid 1996). Several analysts have noticed that these intermediate structures tend to be corporatist and therefore weak as social movements. Personal relations between the leaders play the most important role on every level. The development of social movements and voluntary associations in Estonia during the 1990s went through four stages. Initially, large-scale mass movements emerged with very general objectives. These then broke into smaller and more specific groups. Next, several parallel initiatives arose in the same subject area, which were in more or less intense rivalry and competition one with another. This tendency has often been explained by leaders' thirst for power, as well as by corporatist tendencies in general. Finally, with the general decline of citizen activism, the movements and associations disintegrated altogether.

The development and functioning of Estonia's environmentally oriented social initiatives came under study in the 1990s (see Niit et al. 1997). Environmental problems are good objects for joint action. As a rule, the main difficulties lie not in the problems themselves but in the relationship between people and the environment (which was transformed in Estonia by privatization) as well as in the group and intergroup relations (e.g., within the communities that oppose themselves to the environment and are vehicles for environmental formation or transformation). Degradation of both the natural and man-made environment may become an impetus for the emergence of environmentally oriented groups or movements that aim to realize some common environmental interest or to oppose the rival interests of other groups or movements. In undertaking this study, we tried to encompass the different structures, modes of action, and ideologies of the whole range of Estonia's environmentally oriented initiatives, instead of focusing only on those of the most visible or active. Our approach was not "institution centered" or based on formal parameters of these initiatives. Instead, we tried to study these initiatives from the "inside," relying to a large extent on the reports of the participants themselves. Using participant observation, analysis of documents, and in-depth interviews with movement leaders and activists, we tried to observe the inner dynamics of the movements (their emergence and development, strategies, successes, their relations with the outside society) and their impact on their participants (motives for participating, values, satisfaction).

Environmental Movements

There is a long tradition of nature protection in Estonia. According to a representative survey, by 1982, the Estonian population was already relatively well informed about environmental problems, held pro-environmental attitudes, actually participated or were ready to participate in pro-environmental actions, and were actively concerned about environmental matters (see Lauristin and Firsov 1987). Glasnost enabled authors to publish and openly discuss environmental facts that had been kept secret up to that point. This change produced a rapid increase in environmental knowledge in the general population. At the same time, the mass media actively propagated alternative green values, contributing to the development of general environmental awareness.

In the 1980s, the objective deterioration of their environmental situation became apparent to most, if not all, Estonians, whether from pollution of the air especially in the northeast (due to mining and other heavy industries), of the water in the Baltic Sea, or of the soil all over Estonia (due largely to industrialized agriculture and the Soviet Army). In the 1990s, the (re)privatization of land and rapid economic development threatened previously protected ecosystems.

Environmental movements were among the first mass social movements during perestroika. In 1986–88, discussions of the alternatives in city planning for Tallinn were widely publicized: should a hundred-year-old wooded district in the center of the city be preserved or a new highway built through it?; should an exotic relict tree be preserved or cut down to make way for an opera house?; and should a historic building be preserved or replaced with a badly needed hotel? are just a few of the choices that were widely discussed and reported. Demonstrations were mounted, discussion meetings held, and petition campaigns launched on these and other environmental issues. However, the strongest impetus to organize environmental movements came from Moscow. When the central authorities announced plans for new phosphorite mines in northern Estonia, these were published and widely discussed by specialists, spurring a nationwide campaign against phosphorite mining. Beside its strictly environmental orientation, the anti–phosphorite mining movement was also antitotalitarian and antibureaucratic: it demanded democracy and national revival. On the whole, this environmental movement had positive outcomes: phosphorite-mining plans were denounced, activists were motivated to form new movements and parties, and participants and sympathizers raised their self-esteem.

Environmental movements were among the most influential oppositional forces in 1988–89. They obtained huge support and could mobilize large numbers of people because they were addressing critical problems, whose solution could bring tangible results, and because public opinion was prepared to openly discuss these problems, which were not deemed to be too radical for that time. Examples of environmental movements and voluntary associations which were active during the transition period follow.

National Environmental Movements

The Estonian Green Movement (EGM) was established in April 1988 as a political mass movement at the peak of public environmental activism, during the "War against Phosphorites." Unlike the traditional nature protection movements, it was politically oriented, relatively radical, and promoted a green ideology. Its general democratic and national liberation aims were close to those of the Estonian Popular Front. Among its declared environmental goals were halting the deterioration of Estonia's environmental situation and promoting environmentally sound technologies and healthy lifestyles. Its radical political goal, as declared in 1989, was to establish the Green Republic of Estonia. As a democratic, demilitarized, anti-consumerist, and anti-technocratic, law-governed state, the Green Republic would follow an alternative (nonsocialist and noncapitalist) road of development within a healthy, nuclear-free environment. EGM's programmatic aims included the reasonable usage of natural resources; the preservation of natural and artificial landscape, plant cover, and wildlife typical for Estonia; and the promotion of a healthy, human-friendly man-made environment. To exert political pressure and to send representatives to Parliament, several green parties were founded, which united into the "Estonian Greens" in December 1991.

The Estonian Green Movement operated along nonhierarchical, democratic lines; it maintained relations with its international counterparts (Friends of the Earth, Coalition for a Clean Baltic), with loosely connected local and regional suborganizations, and with independent specialized action groups (mass-media groups, alternative energy groups). As an association of green action groups that were independent of any political parties, including the Estonian Greens, the EGM pursued its aims through publicity, networking, and direct action (such as picketing, demonstrations, plebiscites, nonviolent opposition, discussion meetings, demonstrative happenings). On the local level, groups of greens engaged in concrete activities for improving the environment of their localities; on the regional level, groups addressed larger than local problems and lobbied their commune representatives; and on the national level, EGM organized and coordinated all-republican activities and campaigns and lobbied members of Parliament. The movement published several periodicals on soft agriculture, biodynamics, and vegetarianism, as well as on general problems of the greens.

The movement's level of participation and activism has notably declined since 1988, as all-national causes lost their appeal and as EGM turned to locally oriented more or less radical action. In the 1990s, the Estonian Green Movement was an umbrella organization, a loose network of specialized workgroups and sixteen regional organizations. In May 1993, it had two thousand members. EGM evolved from a mass movement with broad democratic and political aims, to a specialized and professionalized association of green action groups, and finally to a purely environmental organization, losing many of its previous supporters. Small specialized units sometimes acted independently, under the cover of the movement, since they had overlapping membership and the same commitment, ideology, and aims. Some were pure action groups (youth groups, pacifist groups), and others mainly propagated new ideas and values (alternative lifestyle, renewable energy, twenty-first-century, waste management workgroups). Relying on Western financial help and cooperation, many of these groups remain quite active, although less visible than in the times of perestroika. Speaking about the success of the greens, one cannot omit their participation in Parliament (where they hold eight seats as a registered parliamentary party in the government coalition). That the leader of EGM was at the same time Estonia's minister of ecology was variously seen as a great success or as a major mistake of the greens. Since Estonia regained its independence, the greens have lost their political influence. The dilemma—either to remain a pure protest movement or to engage in compromises (such as accepting phosphorite mining for economic reasons)—has lost its immediacy.

Local and Regional Action Groups and Protest Movements

These smaller groups were usually inspired by one specific local environmental problem and initially targeted only this critical issue. They could act on the regional (Virumaa Foundation), local (Green Maardu), city quarter (Kalamaja Society), or neighborhood (Suda Society) level. Some, like the Suda Society, ceased operations after they achieved a certain success; others, such as the Kalamaja Society, evolved into permanently active local societies, widening their scope of interests to all local problems; acting like self-management organizations, they have offered an alternative to local housing maintenance authorities.

The Virumaa Foundation was the first voluntary nonprofit NGO in Estonia. An educational, scientific, and charitable organization, the foundation sought to ensure a humane economic, social, and natural environment in Ida-Virumaa—the location of the "War against Phosphorites" mentioned above and a strongly polluted and demographically unbalanced region. Because Ida-Virumaa is critical to the future of Estonia, the Virumaa Foundation has had nationwide significance. Accordingly, in 1990, it joined the Center of European Foundations. Independent of political parties and barred by its statutes from participating in any election campaigns, the foundation was widely supported by the general population during the Soviet era, when it seemed to fight for the common interests of all Estonians. Gradually, as it became more narrowly oriented to concrete or local problems in Ida-Virumaa, public attention and participation declined.

The "Green Maardu" Association was established in 1988 as a reaction to the highly polluted industrial environment of Maardu, one of Tallinn's suburbs. The association put together an alternative plan for local development, containing an ecological village, newly planted trees in the pits of open-cast mines, an agricultural reserve, and tourist centers, among other innovations. Although the plan was widely advertised, it was supported neither by the public nor by the environmental authorities. On the whole, in the context of the processes of economic and legal restructuring in Estonia, such local initiatives have become rare and relatively ineffective.

Traditional Nature Protection Organizations with Hierarchical Structure
Initiated by intellectuals in 1966, the Estonian Society for Nature Conservation has been the nation's largest NGO, with 23,000 members and forty-nine sections. As a traditional nonpolitical organization with a conservative and humanitarian bent, the society has united a broad range of the public, from schoolchildren and teachers to members of the judiciary, in support of its environmental aims. These aims have been oriented toward traditional nature protection: preserving local cultural traditions, and promoting ecological education, while at the same time claiming essential connections between nature and culture. It has pursued them by organizing summer seminars and people's universities, constructing landscape tracts and study centers of nature protection, conducting na-

ture protection days in schools, and engaging in landscape guardianship and home decoration. Each year, the society organizes all-republican gatherings in a different part of Estonia, thus making the general population aware of local environmental problems. Although it has never directly exerted any political pressure, its leaders have always been part of the official state apparatus; indeed, because one such leader was a high-ranking Communist Party functionary during the phosphorite war, it could take no overtly oppositional stance. The society maintains good relations with the former elite, many of whom are members and whose presidential candidate it supported in the parliamentary elections.

In the 1990s, the Estonian Society for the Nature Conservation presented itself as having been a covertly oppositional organization that had managed to preserve not only Estonian landscapes but also national and cultural identity. Before independence, it had had many members in all parts of Estonia; after a short period of decline during the transition, membership slowly began to rise again. Removed from explicitly political concerns, the society has provided people a rewarding way to escape from their everyday troubles. Because it deals directly with small-scale practical problems, the society has united people with no political or social ambitions; most of the Estonian Greens have been its members in the past.

Nearly all of Estonia's environmentally oriented movements became less active in the 1990s than they had been during the Soviet era. This decline in activity stemmed partly from the restructuring of the whole system of nature utilization and protection, and partly from the temporary improvement of the objective environmental situation (as the result of the abrupt decrease in both agricultural and industrial production). Another significant reason for the decline was that other, more direct opportunities for political action have emerged. People with political ambitions left the environmental movements because they no longer needed to hide those ambitions under the green label.

The interviews we conducted clarified a number of challenges facing Estonian environmental movements. First, nearly all government representatives who had formerly been activists in a particular movement still felt a strong commitment to that movement and preferred to speak on its behalf. Indeed, some members of political parties identified more with their former movements than with their parties, which reflects the strong affiliative power of such voluntary initiatives. Second, because many persons were members of several environmental organizations at the same time or in succession, most activists of different movements are acquainted, and personal sympathies or rivalries play an important role in those movements' dynamics. Third, because most respondents were deeply involved in their respective movements, stereotypes about the "glorious past" or the mythical self-image of a particular movement have clouded their perceptions. Our method of interviewing enabled us to obtain a factual image of the here and now, to penetrate those misperceptions.

Fourth, most movements oriented to more general issues were formed as Estonian counterparts of similar movements in the West. Although this "shadowing" of Western groups has enabled them to gain access to Western informational and material resources, it has also made them vulnerable.

And, fifth, during the struggle for independence, most of the new environmental movements stood in opposition to the old establishment, striving to create something completely new. In the independent Estonia it was often difficult for these groups to maintain their opposition and to continue their protest activities. We noticed that current environmental movements were guided by different underlying values and motives. Thus they used their declared environmental orientation

• to hide their other orientations (Society for Nature Conservation was largely oriented to the preservation of cultural heritage, Estonian Greens had primarily political aims);
• to motivate them in solving real environmental problems, whether through protests, active participation, or both (tenants' societies, local residents' societies, Estonian Fund for Nature);
• to promote informal interaction and self-development (Virumaa Foundation, Homesite Movement).

Due to the political context and the more immediate problems posed by their uncertain legal status, relatively few non-Estonians participate in environmentally oriented movements, although it once seemed that, being politically neutral, these movements might become the basis of their integration (or of the mutual adaptation of Estonian and non-Estonian communities). The Estonian mass media have tried to portray local Russians as an out-group, and to treat them as an "environmental problem." Because other problems have seemed more urgent, there are no special environmental or housing movements consisting only of Russian speakers. People of other nationalities who have integrated into the Estonian society have participated in Estonian-language organizations.

In the 1990s, the general decline of citizen initiatives and striking changes in public opinion priorities changed the landscape of environmental activism. An opinion poll in west-Virumaa has revealed that, in 1993, the majority of local inhabitants preferred phosphorite mining (which would guarantee jobs) to nature conservation (which could bring unemployment). During the Soviet era, nature preservation had a hidden political agenda—to protect the local culture and national heritage. During perestroika in the 1980s, environmental protection appeared as a open political movement (see Barcena et al. 1997; Niit et al. 1997; Yanitski 2000). The "War against Phosphorites" was an environmental protection campaign in the late 1980s, carried out to prevent the opening of new phosphorite mines in northern Estonia, which would have very seriously damaged Estonia's environment and necessitated bringing in large numbers of foreign workers. Environmental protection was an accepted method of expressing fears of yet another wave of migrant labor, Russification, and the destruction of Estonian culture, rather than a movement dealing only with environmental issues. Later, motives of ownership and personal economic gain began to prevail, which undermined the development of collective initiatives.

Since Estonia regained its independence, the popularity of mass environmental movements has steadily declined and, with it, participation in such movements. Indeed, public participation in the country has also declined as a whole. There is moderate-to-low participation in public elections and low participation in voluntary organizations (2 percent of the population are members of political parties, and only 2 percent are active in environmental organizations—compared to nearly 15 percent in the old democracies; COWI 2000). Attitudes toward public participation in Estonia differ from those in western Europe; the right to political participation is not taken for granted, and the majority of the population still see themselves as subjects of the government rather than as participants in the political process. Only a minority of the population sees it as an obligation to take an interest in politics. (COWI 2000). Nevertheless, environmental protection issues are discussed in the media, mostly in the context of economic and social issues. There are no social movements focusing exclusively on environmental justice as it relates to the physical environment. Neither have issues of environmental justice appeared in human rights–related NGOs.[4]

Today, environmental law and environmental justice in Estonia are addressed together with other environmental issues by two main environmental NGOs—the Estonian Green Movement (EGM; http://www .roheline.ee) and the Estonian Fund for Nature (ELF; http://www.elfond .ee). These are the only organizations in Estonia providing legal environmental help to individuals and organizations. EGM tries to promote public participation in environmental decision making by disseminating information on the Aarhus Convention and the principles of participatory democracy. In recent years, ELF has sponsored more and more activities dedicated to environmental education, public awareness, and public participation in environmental decisions and activities.

For instance, starting in 2001, the Estonian Fund for Nature has undertaken several environmental law projects. Since 2003–2006, ELF has handled 118 environmental cases, including thirteen legal cases (Vaarmari 2003; Vilbaste 2003); in 2002–2004, it provided free environmental legal help for environmental NGOs and individuals. The criteria for providing legal help in cases were significant public interest and sufficient grounds to win the case in court. Also, both EGM and ELF have been active in the propagating and implementing the principles of the Aarhus Convention in Estonia (Poltimäe et al. 2004; EKO 2005).

Independent analysis (COWI 2000) shows that public participation in environmental policy making is mediocre in Estonia. There is only moderate interest in environmental information, and 70 percent of the procedures for environmental impact assessments take place without public participation. At the same time, NGO participation in environmental policy documents has increased. Thus the Aarhus Convention plays a wider role in Estonia than protection of the environment by emphasizing the values of participatory democracy and environmental political rights. Implementation of the convention's principles continues to pose a challenge.

The National Strategy on Sustainable Development: Focus on Participation

Approved by Parliament in 2005, the Estonian National Strategy on Sustainable Development (SE 21) sets out a sustainability strategy for the country until the year 2030. Its central aim is "to integrate successful global competition with a sustainable development model and preservation of the traditional values of Estonia" (Ministry of the Environment 2005).

The strategy was drawn up by a consortium under the leadership of Tallinn University in an open and participatory process that involved all key stakeholders and was designed both to produce better cross-sectoral integration and to raise public awareness. In parallel with the work of over fifty experts from different spheres of life assigned to five work-groups, the key aspects of the strategy were comprehensively discussed with social partners, stakeholders, and the public.

Mindful of the interaction between development and environmental protection, the strategy defines four principal preconditions for sustainable long-term development in Estonia:

1. *Preserving Estonian cultural space*   According to the Constitution of the Republic of Estonia, the state of Estonia shall "ensure the preservation of the Estonian nature and culture through the ages." The maintenance of the Estonian nation and culture constitutes the cornerstone of sustainable development of Estonia.

2. *Providing for the welfare of citizens*   Welfare is defined as the satisfaction of the material, social, and cultural needs of individuals, accompanied by opportunities for individual self-realization.

3. *Preserving social coherence and solidarity*   Achievement of the first two preconditions will be possible only if a clear majority of citizens enjoys their benefits and believes in and contributes to their achievement—and only if the price for achieving the preconditions is not destructive of society as an integral organism.

4. *Maintaining ecological balance*   A central precondition for sustainability, maintaining ecological balance is also Estonia's contribution to global development, following the principle that there must be balance both in material cycles and in energy flows at all levels of the living environment.

Estonia's development is judged sustainable if it meets all four preconditions at the same time—viable Estonian language and culture, a high quality of life, an integrated and just society, and a balance between nature and human activity—with measurable outcomes. Sustainability is endangered if any of these major preconditions is neglected. To achieve them, the strategy strongly emphasizes a more socially balanced path of development, shifting toward greater public participation and more equal access to the country's resources, both economic and environmental.

Conclusion

During the post-Soviet era, Estonia has gone through radical socioeconomic transformations, which have produced controversial results. On
the one hand, the country has achieved rapid economic growth; on the
other, economic development has been accompanied by the aggravation
of several social problems. The industrial northeastern region of Ida-
Virumaa is the area of most critical concern both socially and environmentally; its inhabitants (over 60 percent of whom are non-Estonians)
are more exposed to environmental degradation and harmful conditions
and have fewer opportunities to participate in social action than those
living elsewhere in the country. Despite objective and perceived social injustice, participation in protest movements is relatively low in Estonia.
On the positive side, however, green voluntary organizations are taking
up environmental justice issues, and the National Strategy on Sustainable
Development is also oriented toward the increase of social and environmental justice and public participation.

Notes

1. Moreover, the state of public health in Estonia has deteriorated in a number
of other ways: for example, Estonia's incidence of HIV/AIDS is the highest in the
European Union (see Järve et al. 2006).

2. The Gini index provides a measure of income or resource inequality within a
population. A low Gini index indicates more equal income distribution, while a
high Gini index indicates more unequal distribution.

3. According to WWF report, Estonia is among five European countries with the
largest ecological footprint per person.

4. Legal Information Center for Human Rights (LICHR): www.lichr.ee.

References

Aarelaid, A. 1996. "Civic Initiative and Human Development." In P. Järve, K.
Toomel, and L. Viik, eds., Estonian Human Development Report 1996, 46–50.
Tallinn: United Nations Development Program.

Barcena, I., P. Ibarra, and M. Zubiaga. 1997. "The Evolution of Relationship be-
tween Ecologism and Nationalism." In M. Redclift and G. Woodgate, eds., The
International Handbook of Environmental Sociology, 300–315. Cheltenham,
UK: Elgar.

Consultation Engineers and Planners, AS (COWI). 2000. "Attitudes and Barriers to Public Participation in Environmental Decision Making in Estonia." Discussion paper. Kongens Lyngby, Denmark.

Estonian Council of Environmental NGOs (EKO). 2005. "NGO Report about the Implementation of the Aarhus Convention in Estonia for the Second Conference of the Parties." Almaty, Kazakhstan, May 25–27. http://www.eko.org.ee/.

Hallik, K. 2002. Nationalising policies and integration challenges. In M. Lauristin and M. Heidmets, eds., *The Challenge of the Russian Minority: Emerging Multicultural Democracy in Estonia*, 65–88. Tartu: Tartu University Press.

Hallik, K., Poleschuk, V., Saar, A. and Semjonov, A. 2006. *Estonia: Interethnic Relations and the Issue of Discrimination in Tallinn*. Tallinn: Legal Information Center for Human Rights.

Heidmets, M., ed. 2007. *Estonian Human Development Report 2006*. Tallinn: Public Understanding Foundation.

Kiivet, R., and J. Harro, eds. 2002. *Health in Estonia 1991–2000*. Tartu: Estonian Health Insurance Fund.

Kreitzberg, M., A. Valtin, and M. Fedina. 2006. "Household Living Niveau 2005." In S. Linnas, ed., *Estonian Social Survey*. Tallinn: Social Statistics Department of Statistics Estonia.

Kruusvall, J. 2002. "Social Perception and Individual Resources in the Integration Process." In M. Lauristin and M. Heidmets, eds., *The Challenge of the Russian Minority: Emerging Multicultural Democracy in Estonia*, 117–162. Tartu: Tartu University Press.

Lauristin, M. 2003. "Social Contradictions Shadowing Estonia's 'Success Story.'" *Demokratizatsiya*. http://www.findarticles.com/p/articles/mi_qa3996/is_200310/ai_n9310188/.

Lauristin, M. 2004. "Hinnangud Eesti ühiskonnas toimunud muutustele [Attitudes towards social changes in Estonia]." In V. Kalmus, M. Lauristin, and P. Pruulmann-Vengerfeldt, eds., *Eesti elavik 21.sajandi algul* [Estonian life-world in the beginning of the 21st century], 231–249. Tartu: TÜ Kirjastus

Lauristin, M., and B. Firsov, eds. 1987. *Massovaja kommunikatsija i ohrana sredõ*. [Mass Communication and Environmental Protection] Tallinn: Eesti Raamat.

Lauristin, M., and R. Vetik, eds. 2000. *Integration of Estonian Society: Monitoring 2000*. Tallinn: Integration Foundation.

Meikar, T., ed. 2005. *Metsaseadustest Eestis* [On Forestry Legislation in Estonia]. Tartu: Eesti Metsa Selts.

Merisaar, M. 2007. "Estonian Environmental NGOs on the EU Environmental Policies."

Ministry of the Environment. 2005. *National Strategy on Sustainable Development: Sustainable Estonia 21*. Translated by M. Maran. Tallinn.

Niit, T., M. Raudsepp, and K. Liik. 1997. "Housing and Environmental Movements in Estonia." In K. Lang-Pickvance, N. Manning, and C. G. Pickvance, eds., *Housing and Environmental Movements: Grassroots Experience in Hungary, Estonia and Russia*, 47–60. Aldershot, UK: Avebury.

Poltimäe, H., P. Kuldna, M. Merisaar, and T. Kolk. 2004. *Keskkonnainfo kättesaadavusest ja otsustamises osalemise võimalustest Eestis* [On the accessibility of environmental information and possibilities to participate in decision-making in Estonia] SEI väljaanne nr. 6 [Stockholm Environment Institute Report no 6]. Tallinn: Säästva Eesti Instituut.

Ranniku, V., ed. 1996. *Igaüheõigus* [Right to free access]. Tallinn. Keskkonnaministeerium.

Raudsepp, M. 2005. "Eestlaste loodusesuhe keskkonnapsühholoogia pilgu läbi [Estonians' Relation to Nature from the Environmental Psychology Viewpoint]." In T. Maran and K. Tüür, eds., *Eesti looduskultuur* [Estonian nature culture], 379–420. Tartu: Eesti Kirjandusmuuseum.

Raudsepp, M. 2007. Report on the nationwide surveys "Environment and Us" (2002) and "Me, Nature and Forest" (2006). Available at Environmental Psychology Research Centre, Tallinn University.

Saar, E. and Lõhmus, K. 2003. Sotsiaalne ebavõrdsus Eestis: tegelikkus ja inimeste hinnang (Social inequality in Estonia: reality and subjective assessment). In R. Vetik, ed. Estonian Human Development Report 2003, 52–57. Tallinn: Institute of International and Social Studies at Tallinn University.

Vaarmari, K. 2003. "Ülevaade ELFi keskkonaõiguseabi projekti tööst [Overview of the environmental legal aid project in Estonian Fund for Nature]" *Roheline Värav* [Green Gate], November 13.

Vilbaste, K., ed. 2003. *Keskkonna õigus: Eestis õigusabi vajanud juhtumid* [Environmental justice: Estonian cases that required legal help]. Tartu: Estonian Fund for Nature.

Yanitsky, O. 1985. "Human Relations, Environment and Urban Ecology: Reproductive Approach." In T. Niit, M. Heidmets, and J. Kruusvall, eds., *The Socio-Psychological Basis of Environmental Design*, 30–34. Tallinn: Estonian Branch of Soviet Psychological Society and Tallinn Pedagogical University.

Yanitsky, O. 2000. *Russian Greens in a Risk Society: A Structural Analysis*. Helsinki: Kikimora.

10

Environmental Injustices, Unsustainable Livelihoods, and Conflict: Natural Capital Inaccessibility and Loss among Rural Households in Tajikistan

Dominic Stucker

Environmental injustice—in the form of limited access to and control of natural capital—negatively impacts the livelihoods and security of rural households in Tajikistan.[1] Its primary victims are impoverished households, especially those headed by single women, living in Gorno-Badakhshan, and/or engaged in cotton farming (see figure 10.1).[2] Moreover, the country's governance structures and processes remain inadequate, particularly at the highly centralized national level, and often perpetuate environmental injustice and conflict. Indeed, civil society in Tajikistan possesses very limited influence, and public participation in decision-making processes is the exception.

This chapter offers a working definition of environmental justice in Tajikistan and then supports the above thesis by (1) describing the rural vulnerability context; (2) highlighting unsustainable—and dangerous— livelihood strategies, including child labor, outmigration, human trafficking, drug trade, and militant fundamentalism; (3) analyzing the access to natural capital and the processes of natural capital loss among different types of rural households; (4) analyzing the shortcomings of environmental governance, especially at the national level; and (5) discussing the state of an environmental justice movement in Tajikistan and possible steps forward. My research was carried out using Sustainable Livelihoods Framework analysis (Chambers and Conway 1991; DFID 1999) focused on natural capital, the essential and non-substitutable livelihood asset for rural, agrarian households (Neefjes 2000, 89). Development workers often employ this tool in the field to help them first identify obstacles to sustainable livelihoods and then design their programs accordingly. (See figure 10.2.) The framework includes the shocks, trends, and seasonality of the vulnerability context; five forms of capital that households can draw upon; mediating governance structures and processes;

Figure 10.1
Bakhrom-aka, rural farmer, shows off his wheat harvest. Photo taken by the author in 2001.

existing livelihood strategies; and, finally, livelihood outcomes, which can alter the vulnerability context (see Carney et al. 1999 for a review of Sustainable Livelihood Frameworks).

As Peace Corps Volunteers, educators, human rights advocates, and environmentalists, my wife, Abigail, and I lived in rural, subsistence communities in Tajikistan, Uzbekistan, and Kyrgyzstan for over three years. We formed lasting friendships with neighbors and colleagues and witnessed firsthand many of the daily struggles that confront rural households. We were moved by the resourcefulness and tenacity of community members, by the hospitality and many kindnesses they extended to us, and by the tragedies that befell them when economic pressures were too great. Family members migrated abroad, parents sent children

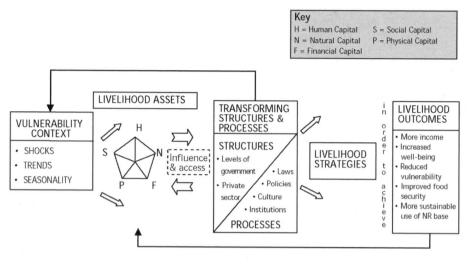

Figure 10.2
Sustainable Livelihoods Framework (DFID 1999, 2.1).

they could not feed to orphanages, daughters were lost to human trafficking and sons to militant fundamentalist groups. Our experiences among these families inspired my research.[3]

A Working Definition of Environmental Justice in Tajikistan

The Central Asian country of Tajikistan is overwhelmingly rural, agrarian, and impoverished, highly dependent on natural capital (see note 2, figure 10.3). Even though two-thirds of all households depend on agriculture—both farming and raising livestock—for their livelihoods, only 5 percent of the land is arable (UNECE 2004, 15).

Access to and control of natural capital are severely limited for rural households, threatening the sustainability of their livelihoods. With nearly 70 percent of its people living below the poverty line (UNRC 2006), and only now recovering from a protracted, bloody civil war in the 1990s, Tajikistan is the poorest country in the former Soviet Union (UNECE 2004, 12). Its unsustainable rural livelihoods and attendant poverty only serve to increase social conflict and the likelihood of renewed violent conflict (Stucker 2006).

The definition of environmental justice can vary by place and change over time (Holifield 2001, 86). Unlike conventional definitions, mine

Figure 10.3
Political map of Tajikistan (UN 2004).

focuses on environmental degradation instead of environmental hazards. Though landslides, earthquakes, abandoned dumps of toxic tailings, and pollution from aluminum and concrete factories compromise the security and health of hundreds of families in Tajikistan (de Martino et al. 2005; UNECE 2004, 17; Carius et al. 2003),[4] the loss of natural capital impacts the entire rural population, especially the most vulnerable households. My working definition attempts to reflect the current realities and priorities of rural Tajikistan:

Environmental justice is manifest when every household's right to access and capacity to manage the natural capital necessary to sustain the livelihoods of its members are secured and guaranteed.

This definition is based on the conviction that individuals should have the right both to a sustainable livelihood (see ECC 2000, 3b, 9b) and to environmental security at the household level. It represents the bare minimum required to alleviate poverty and minimize the potential for social and violent conflict. If this degree of environmental justice can be secured, the definition can—and should—be expanded to include additional rights and responsibilities, ensuring health and security for all life, ecosystems, and future generations.

Regardless of the scope of the definition, essential prerequisites for guaranteeing environmental justice include open access to information for women and men and public participation in decision-making processes. Indeed, even though "there exists no clear and universally accepted definition of environmental justice...some notion of effective citizen participation and empowerment underlies everyone's conception of it" (Foreman 1998, 34). In this regard, the national government and local civil society must play key roles in stimulating a robust environmental justice movement in Tajikistan.

The Rural Vulnerability Context

Historical Shocks

The collapse of the Soviet Union in 1991 was followed by an economic crisis and loss of social services in Tajikistan. Leading up to independence, Tajikistan had been the poorest and most highly subsidized republic within the Soviet Union, receiving almost half of its revenues from Moscow (UNECE 2004, 4). Not only did these subsidies abruptly stop, but Tajikistan's trade, heavily dependent on former Soviet republics, also crumbled. Lack of state financial resources was compounded by a brain drain of intellectuals and managers, who emigrated abroad for work. Health care, education, transportation, and irrigation systems began to deteriorate. A lack of state social services heightened Tajiks' dependence on subnational regional and clan networks, which were a significant factor leading up to the civil war, and which became deeply entrenched as a result (Olimova 2004; Gretsky 1995).

The civil war dragged on from 1992 to 1997, further devastating the Tajik economy and society. Out of a total population of 5.1 million people in 1989, an estimated 50,000 to 100,000 people were killed, hundreds of thousands were internally displaced, and close to 600,000 emigrated (UNECE 2004, 6, 3). In 2003, after seven years of gradual improvement from its low in 1996, Tajikistan's GDP per capita had not even reached half of its already impoverished 1990 level (UNECE 2004, 6).

In 1999, just two years after the signing of the Peace Accords, Tajikistan was beset by three years of devastating drought, resulting in the extensive loss of natural capital and prolonged food insecurity.[5] Crops failed, yields plummeted, and arable land was lost. Given global trends of climate change, including higher temperatures and loss of precipitation in some regions, the shock of drought is likely to recur in Tajikistan (Breu and Hurni 2003, 9; Solomina 2005, 70).[6]

These shocks, both historical and recurring, have severely impacted the population and development of Tajikstan. What is more, they have often set harmful trends in motion.

Current Trends
Although the 1997 Peace Accords helped bring an end to widespread violent conflict in Tajikistan, hopes for improved governance and, by extension, reduced economic vulnerability, have been left unfulfilled. Imomali Rahmon, the president of Tajikistan since 1992, heads a highly centralized and authoritarian national government that is becoming less and less democratic. The runaway victories for the president's party in the 2005 parliamentary elections and for the president himself in the 2006 elections have been widely criticized by international and national experts (OSCE 2006b; Lillis 2006). Indeed, in a recent survey measuring democratic governance (a composite figure of accountability and public voice; civil liberties; rule of law; and anticorruption and transparency), Tajikistan ranked lowest among those countries surveyed in the Caucasus and Central Asia (Freedom House 2007). According to the same study, Rahmon maintains "dominance in national politics by undercutting and dividing the political opposition, driving out international nongovernmental organizations and tightly controlling local ones, harassing human rights organizations, marginalizing independent media, and strengthening his political party's majority in the legislature" (Freedom House 2007). Such activities severely compromise the emergence of environmental governance and impede the formation of an environmental justice movement.

Moreover, Tajikistan has the highest fertility rate in the former Soviet Union, with attendant population growth increasing demand on already scarce natural capital, especially in rural areas. Even after steady declines since 1990, Tajikistan's fertility rate was still at 3.1 live births per woman in 2001 (WHO 2006a). Given that internal migration is predominantly from urban to rural areas (UNECE 2004, 15), small towns and villages feel the greatest degree of population pressure. Indeed, "Tajikistan's rural settlements are facing the highest demographic increase among the countries of the [former Soviet Union]" (UNECE 2004, 15), helping maintain the overwhelmingly rural population at 70 percent (UNRC 2006).

The country's large demographic youth bulge poses serious challenges to education, negatively impacting young women in particular. The population under fifteen years of age has remained at the steady 40 to 45 percent of the total it represented from 1990 to 2002 (UNECE 2004, 3). A recent study showed that the number of children enrolled in compulsory education is dropping (Falkingham and ADB 2000, 65); even those who are enrolled miss months of school while harvesting cotton, described in more detail below. Furthermore, the gap between enrollment of boys and girls in education is widening. The graph in figure 10.4 represents the education situation in 1998; it has continued to worsen ever since (Falkingham and ADB 2000, 65). Whereas women made up 58 percent of students in higher education in 1990, they represented only 34 percent in 1998 (Falkingham and ADB 2000, 65).

As young people grow into adulthood, high unemployment ensues. Unemployment of males aged sixteen to twenty-four has been over 50 percent, making them more susceptible to recruitment by militant fundamentalist groups (De Nuebourg and Namazie 1999 in Falkingham and ADB 2000, 44). Young women's unemployment is nearly as high and

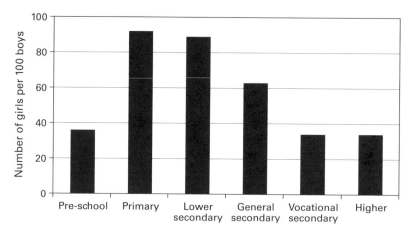

Figure 10.4
The gender gap in education in Tajikistan, 1998 (Falkingham and ADB 2000, 65).

Figure 10.5
Photo of women, girls, and boys in rural Tajikistan (WB et al. 2005).

compounded by their working largely in the least profitable sectors of agriculture, education, and health care (Falkingham and ADB 2000, 40). Indeed, in 2006, women made up some 70 percent of the agricultural workforce (UNRC 2006). In the domestic sphere, rural women continue to bear the greatest share of the unpaid labor and enjoy the smallest share of leisure time (see Falkingham and ADB 2000, 51). (See figure 10.5.)

Thus, it is clear that poverty in Tajikistan is chronic and predominantly rural (see also UNECE 2004, 13). The country was one of the twenty poorest in the world in 2003 (ICG 2003) and had a GDP per capita of $287 in 2004 (UNRC 2006), or less than 80 cents per person per day. (See figure 10.6.)

Poverty-related malnutrition and diseases such as typhoid, malaria, and dysentery are on the rise in Tajikistan (WB 2005, x), reflected in the highest child mortality rates of the former Soviet Union, especially among the rural population (WB et al. 2005). In 2004, 120 out of every 1,000 boys and 115 out of every 1,000 girls died before their fifth birthday (WHO 2006b). Pathogens from contaminated water sources, airborne particulates from burning traditional fuels indoors, and poor sanitation contribute to this increase in child mortality.

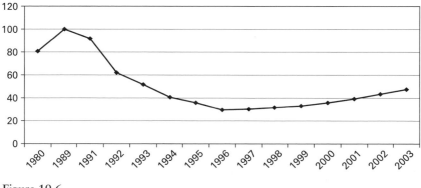

Figure 10.6
Real GDP development of Tajikistan (1989 = 100; UNECE 2004, 12).

In sum, the above shocks and trends pose significant daily challenges to rural households. The national Rahmon regime restricts independent media, minimizes public participation in decision-making processes, renders the judicial system largely inaccessible and corrupt, and disempowers civil society. A declining education system, a widening gender gap in schools and universities, and an increase in disease compromise social and human capital. Population growth places additional burdens on the environment, while poverty drives some rural households to engage in livelihood practices that threaten natural capital and their own security.

Unsustainable Rural Livelihood Strategies and Conflict

Many vulnerable households in Tajikistan are forced to engage in livelihood strategies that are unsustainable, dangerous, illegal, and often closely linked to social and violent conflict. This section highlights the most problematic of these strategies, emphasizing the urgent need to address a key underlying cause, environmental injustice.

Child Labor

Though officially condemned by the government, the use of child labor, especially during the cotton harvest, is still widespread (Oxfam 2006). Based on a recent survey conducted in the cotton-farming oblast of Khatlon, children miss up to one-third of their school year (380 academic hours) harvesting cotton (Oxfam 2006, 1), which deprives them of the human and social capital crucial for sustainable livelihoods.

Figure 10.7
Photo of girl harvesting cotton in Khatlon Oblast, Tajikistan (Oxfam 2006a).

Children, especially girls, generally start working in the cotton fields at twelve years of age, and many continue to work there for the rest of their lives (Oxfam 2006, 2). While laboring on state cotton farms, children are generally not provided with food and do not have access to medical facilities, resulting in high rates of influenza, diarrhea, urinary tract and kidney infection, and tuberculosis (Oxfam 2006, 2). Of those parents surveyed, 70 percent reported that their children's health suffered from working under such harsh conditions (Oxfam 2006, 1). (See figure 10.7.)

Children harvest approximately 40 percent of Tajikistan's cotton (Oxfam 2006, 1), but receive only a pittance for their efforts. On average, each child harvests over 500 kilograms (1,100 pounds) of cotton per season. Even though their meager wages of 3 cents/kilogram (1.4 cents/pound) would amount to less than $16 for months of toil, they receive, at best, half that amount after deductions for taxes and transportation (Oxfam 2006, 2). Some children are paid nothing, being compensated instead with permission to collect dried cotton stalks, which can be sold or used for fuel (Oxfam 2006, 2). Cotton-farming households are discussed in greater depth below.

Outmigration
Approximately one million men migrate out of the country each year, mostly working as underpaid construction laborers and market vendors in Russia, with 10 percent of them, or 100,000, remaining abroad indef-

initely (*Eurasia Insight* 2006).[7] Because it decreases the numbers of unemployed young men in Tajikistan and supplements families' incomes with remittances from abroad, outmigration is often viewed as mitigating social conflict (Cincotta et al. 2003). However, in the case of Tajikistan, those who choose to remain abroad are generally the most highly educated and least conservative (UNECE 2004, 13). Given a total population estimated in mid-2008 to be 7 million, the loss of 100,000 educated and skilled individuals each year is more likely to perpetuate than mitigate rural poverty.

Outmigration is, therefore, more appropriately viewed as an indicator of the inaccessibility of livelihood assets and the lack of sustainable livelihood options in Tajikistan. As such, it could foreshadow further declines in human capital akin to the brain drain that occurred following independence.

Human Trafficking

Among the most poignant examples of unsustainable livelihood strategies and social conflict are human trafficking and prostitution. Rural women, already in desperate economic straits, are all too willing to trust acquaintances who promise them legitimate work abroad. To their detriment, women are often trafficked to Russia or the Gulf States as domestic servants or prostitutes (IOM 2001). Although precise data are difficult to collect on the trafficking of women and children, the latter are also being abducted and sexually exploited (IOM 2001).

Drug Trade

Tajikistan's key role in the Afghan opium trade means that drug production, trafficking, and abuse have spread throughout the country (Makarenko 2000; Cornell and Spector 2002). In 2003, the harvest from 1 hectare (2.5 acres) of wheat yielded a rural family approximately $222, whereas the same-sized plot, cultivated with opium poppies, yielded $12,700, or more than 57 times as much (Gerstle 2004). It is no wonder, then, that poppy cultivation continues to spread. In terms of drug trafficking, women generally find it easier to pass through border posts and are thus frequently used as drug couriers or "mules" (UNECE 2004). It is also no wonder that intravenous drug use and attendant HIV/AIDS infections are on the rise. There are an estimated 55,000 intravenous drug users in Tajikistan, a number projected to increase by 10,000 each year (UNOCHA 2006).

There is also a strong link between the drug trade and armed conflict. During the civil war and beyond, drug money was used to finance the activities of all parties involved (see Azamova 2001; Makarenko 2000; Rashid 2000).

Militant Fundamentalism

Tajikistan's burgeoning numbers of disenfranchised young men are particularly susceptible to accepting the U.S. dollars offered by militant Islamic groups in exchange for their allegiance (Rashid 1994; NIC 2004; Baran et al. 2006). For many in Central Asia, militant Islam is perceived as the only form of organized opposition against repressive governments (McGlinchey 2005). Those not recruited by such groups might still support their activities out of fear, desperation, or fundamentalist convictions. Given its long, porous border with Afghanistan and its mountainous terrain, Tajikistan has historically provided militant groups a safe haven as well as transit to and from the fertile, relatively conservative Ferghana Valley, predominantly in Uzbekistan and encircled by Kyrgyzstan and Tajikistan (ICG 2001; Rashid 2000).

The loss of natural capital and compromised livelihood security, though certainly not the only contributing factors, increase the likelihood of people engaging in unsustainable and illegal activities and expressing their grievances through violence (Stucker 2006). If environmental injustices are not rectified, members of poor, rural households will furnish the next generation of child laborers, trafficked persons, drug cultivators, couriers, and addicts, and fundamentalist militants. If nothing is done, most rural households will continue to live out their days in grinding poverty, threatened by disease and economic insecurity, and, quite possibly, by renewed violent conflict (Stucker 2006).

Natural Capital Inaccessibility and Loss among Rural Households

The likeliest victims of environmental injustice in Tajikistan—most often at the hands of their own national government—are impoverished households, those headed by single women, those living in Gorno-Badakhshan, and those engaged in cotton farming. The latter three groups tend to be the poorest, though they also lack access to natural capital independently of being poor. The research presented here clearly indicates that households in these categories have the most limited access to land, water, and livestock, the most important forms of natural capital in rural Tajikistan.

Because studies of natural capital inaccessibility and loss in Tajikistan have not been disaggregated along ethnic lines, no conclusions can be drawn about ethnic-based environmental injustice. In this regard, however, subnational regional identity (such as being from Gorno-Badakhshan) is sometimes reinforced by ethnic identity, though in many other cases, the regional identity eclipses ethnic identity in importance (Olimova 2004). Regional-based environmental injustice may thus be more prevalent than ethnic-based injustice.

Impoverished Rural Households

Impoverished households in Tajikistan, in proportion to their degree of poverty, suffer from limited access to and control of land, water, and livestock. These households are overwhelmingly in rural areas. According to a 2002 survey, only "5.1 percent of the urban population falls into the bottom two wealth quintiles and nearly 60 percent into the wealthiest category [whereas] in rural areas...52.9 percent is in the bottom two [wealth quintiles] and only 5.4 percent in the wealthiest category" (UNECE 2004, 13).

Please note that the following observations about land, water, and livestock holdings among the rural population compare relative poverty within a context of absolute poverty. That is to say, even most of the wealthier rural households struggle to ensure their own livelihoods. Moreover, although households also draw income from nonagricultural sources (Ellis 1998), agricultural ones are of much greater importance in rural areas (MSDSP 2004, 13). For this reason, a sufficient combination of land, water, and livestock is essential to ensure livelihood security for rural households.

Poverty and Land   In general, there exists a direct correlation between the severity of a household's poverty and its inaccessibility to arable land. In Gorno-Badakhshan, for example, 11 percent of the poorest quartile, 2 percent of the second poorest quartile, and 1 percent of the third have no access to land whatsoever (MSDSP 2004, 22). Furthermore, the poorest quartile holds the smallest plots of land, on average a meager 0.30 hectare (0.74 acre) per household (MSDSP 2004, 22). With an average household size of 6.8 persons (MSDSP 2004, 18), this leaves less than 0.05 hectare (0.12 acre) per person, or only 10 percent of the plot size recognized as the food security threshold for an individual— 0.50 hectare (1.24 acres; Scherr 1999).[8] According to recent studies conducted in Khatlon Oblast, households who do secure the right to land

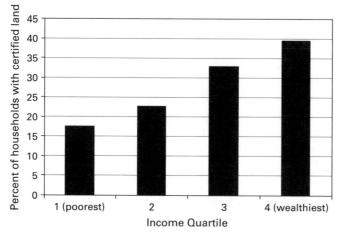

Figure 10.8
Percentage of households with certified land by income quartile in Khatlon Oblast, Tajikistan, 2004 (MSDSP 2005, 12).

usage ("certified land") are most often wealthier and connected with local government officials. They also tend to receive the largest and most fertile plots (ICG 2005, 8; MSDSP 2005, 9). (See figure 10.8.)

Poor soil fertility, contributing to lower crop yields, has also been shown to correlate with the level of household poverty. In this example from Gorno-Badakhshan, shown in table 10.1, yields of potatoes among the poorest households are 20 percent smaller than those among the wealthiest.

Taking into account the smaller plot size and lower yields, the poorest farmers only manage to produce one potato for every three produced by the wealthiest farmers. Such discrepancies are due, in large part, to environmental injustices concerning access to arable land and water, farming extension services, high-quality seeds, fertilizers, and crop storage facilities.

Poverty and Water   An individual needs between 20 and 50 liters (4 and 13 gallons) of uncontaminated water each day to ensure the basic needs of drinking, cooking, bathing, and disposing of excrement (UNWWAP 2003). Even granted that pit toilets, which require no water, are ubiquitous, the amounts displayed for well over half of respondents in figure 10.9 are still far too low to ensure proper nourishment and sanitation for rural households.

Table 10.1
Potato production in Gorno-Badakhshan (GBAO) by income quartile, 2004

Income quartile	Number of households responding	Average area (hectares per household)	Average production (kilograms per household)	Average yield (kilograms per hectare)
1	109	0.05	375	12,979
2	142	0.1	632	14,146
3	150	0.09	727	15,968
4	143	0.1	1,106	16,475
Total	544	0.09	731	15,027

Source: MSDSP 2004, 30.

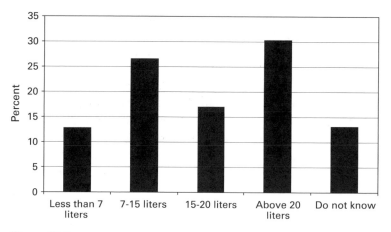

Figure 10.9
Rural domestic water use per capita per day in Tajikistan, 2003 (ECHO 2003 in UNECE 2004, 110).

If those who responded that they "do not know" their daily water consumption were proportionally distributed over the other four group-ings, only 36 percent of rural individuals use more than 20 liters (4 gallons) per day, the lowest threshold for water security, sustenance, and health. This amount would be even lower, of course, if contami-nated water were deducted from the total.

Poverty and Livestock   Since Soviet times, livestock production in Taji-kistan has sharply declined. Pastoral extension services have crumbled

and trained professionals, including veterinarians, are lacking (Morgounov and Zuidema 2001). Combined with the concentration of herds
around settlements, this means that animals are at a heightened risk of
becoming diseased and dying. From 1990 through the end of the civil
war in 1997, livestock production dropped to 15 percent of what it was
in the 1980s (UNECE 2004, 140). State herds were decimated, with 85
percent of livestock now scattered in small, private herds (UNECE 2004,
141). Since 1997, herds have begun to recover, though, as of 2004, they
were only at 30 percent of 1980s levels (UNECE 2004, 141). Furthermore, there is a strong correlation between wealth and the ownership of
livestock among rural households. (See table 10.2.)

Though some of the differences in livestock ownership in this study
from Gorno-Badakhshan are insignificant, households in the wealthiest
quartile, on average, own the most cattle, sheep, goats, chickens, turkeys,
and yaks, whereas the poorest quartile owns the fewest of almost every
type of animal (MSDSP 2004, 24). Like those in Gorno-Badakhshan, the
poorest farmers in Khatlon Oblast can rarely afford to purchase large
livestock, such as cattle (MSDSP 2005, 44).

As table 10.3 indicates, cattle sales generate the greatest average annual cash income, amounting to $228.[9] At approximately $190 per
animal, this is much too expensive for poor households to afford. Such
sales generally benefit the wealthiest farmers; a fraction of the other
households sell goats or sheep for much smaller returns. Such trends, of
course, function to widen the gap between relatively wealthy and poor
farming households.

In sum, arable land, clean water, and healthy livestock are essential,
yet scarce, forms of natural capital for livestock-raising and farming
households. Inadequately small plots of land hinder the ability of rural
households to ensure food security; inadequate water contributes to the
occurrence of illness and decreased land productivity. Finally, small private herds, though they provide important services and products to rural
households, are greatly depleted and prone to disease. The national government, with assistance from the international community and local
civil society, needs to eradicate poverty in Tajikistan, with special
attention directed to securing rural households' access to and control of
natural capital. In other words, the national government must rectify environment injustices if people are to extricate themselves from poverty.

The next three subsections highlight variables independent of impoverishment for explaining environmental injustice among households

Table 10.2
Average number of livestock per household in Gorno-Badakhshan (GBAO) by income quartile, 2004

Income quartile	Number of households in quartile	Cattle	Sheep	Goats	Chickens	Beehives	Turkeys	Donkeys	Rabbits	Yaks
1	174	0.6	1.8	1.7	1.1	0	0.2	0.1	0	0.2
2	174	1.8	2.8	2.8	1.8	0.1	0.1	0.4	0.1	0.4
3	174	2.3	3.6	3.5	2.4	0.2	0.2	0.2	0.1	0.3
4	174	2.5	4.7	3.8	4.2	0.7	0.7	0.2	0.1	0.5
Total	696	1.8	3.2	2.9	2.4	0.3	0.3	0.2	0.1	0.3

Source: MSDSP 2004, 24.

Table 10.3
Annual livestock sales and cash income for households in Khatlon Oblast, 2005

Type of livestock	Percentage of households selling	Average number of animals sold per household	Average income from sales (somoni/dollars)
Cattle	21	1.2	683/228
Goats	20	2.2	205/68
Sheep	12	1.5	181/60
Chickens	4	n/a	21.5/7

Source: MSDSP 2005, 45.

headed by single women, households in Gorno-Badakhshan, and cotton-farming households.

Households Headed by Single Women
As a result of male combatant deaths, households headed by single women began to appear in large numbers during the civil war. Following the war, their numbers have continued to increase due to outmigration of men. At least 5 percent of households are headed by single women in nearly all of Tajikistan's *raiony*, with more than 20 percent in some (WB et al. 2005).

Households headed by single women are very likely to be poor and suffer from attendant environmental injustices concerning land, water, and livestock holding. They are 30 percent more likely to be impoverished than other households (WB 2000), situating most of them in the lowest income quartile of the rural population, as described above.

Households Headed by Single Women and Land   Independent of such poverty, women are especially disempowered concerning land tenure. Though they constitute 70 percent of the agricultural workforce (UNRC 2006), women officially administer only 4 percent of Tajikistan's arable land, and even then, the real power often belongs to a male relative. A local NGO estimated that women, in reality, directly administer only 1 percent of farmland (ICG 2005, 17). This imbalance clearly needs to be corrected by national government land tenure policy (discussed in greater depth under "Cotton-Farming Households" below).

Given the entrenched and pervasive gender inequities that disadvantage most Tajik women, households headed by single women are

Table 10.4
Poverty rates and share of poor by oblast/city, Tajikistan, 2003

Oblast/City	Population	Overall poverty rate (%)	Share of poor (%)
Gorno-Badakhshan	197,000	84	4
Sugd	2,123,000	64	32
Khatlon	2,169,000	78	40
Raiony of Republican Subordination	1,553,000	45	17
Dushanbe	630,000	49	7
Total	6,672,000	64	100

Source: UNDP 2003 in WB 2005.

in a particularly vulnerable situation. The understanding of environmental injustice among this population, however, would benefit from studies disaggregating such households' access to water and livestock holdings.

Households in Gorno-Badakhshan

Though national figures vary according to how one defines and measures poverty, Gorno-Badakhshan is, overall, the poorest oblast in Tajikistan. Table 10.4 allows for comparison between oblasty.[10] Based on this data, rural households in Gorno-Badakhshan are more likely to be impoverished and suffer from greater inaccessibility to land, water, and livestock than those in other oblasty.

Furthermore, independent of such poverty, environmental injustice in Gorno-Badakhshan can be attributed to subnational regionalism, or clan politics, an entrenched form of structural violence in Tajikistan (Olimova 2004; Gretsky 1995; Khudonazar 1995). Regional identities are pronounced and people from different oblasty are often unable to understand one another's speech. Mountainous geography, Soviet and post-Soviet politics, ethnicity, religion, and culture all play a part in defining regionalism. Historically, the importance of regional identities wanes and waxes (Khudonazar 1995), but currently "the collapse of [Soviet] systems of social service, education, health care and social protection has made people reliant mainly on family, relatives and their ethno-regional groups" (Olimova 2004, 98). Regional biases permeate not only government structures but also military, business, and educational institutions, the health care system, and marriage (Olimova 2004, 98).

Indeed, since independence, Gorno-Badakhshan has been virtually ignored by the central government. This mountainous oblast is remote and largely inaccessible from the capital, Dushanbe. As outlined below, state services and infrastructure important to rural livelihoods are lacking, resulting in various forms of environmental injustice.

Gorno-Badakhshan and Land   Geography alone poses daunting challenges to rural households of this oblast. Mountain valleys and plateaus are at altitudes between 3,000 and 4,000 meters (9,850 and 13,100 feet) and arable land constitutes only 0.4 percent of the oblast's total area, limiting crop production (Breu and Hurni 2003, 8). Because the steep valleys and short growing season make farming even more challenging, raising livestock is of greater importance for rural livelihoods.

Processes of natural capital loss in Gorno-Badakhshan, such as deforestation, erosion, and loss of soil fertility, are attributable, in part, to regionalism and attendant state neglect. By failing to provide adequate electricity and natural gas infrastructure in Gorno-Badakhshan (WB et al. 2005), the state has compelled rural households to rely on traditional sources of fuel, especially wood, shrubs, and manure. Compounded by the oblast's relatively high population growth and the absence of Soviet fossil fuel subsidies (Zibung 2003, 35), seasonal deforestation continues, especially during the long, cold winters.[11] Less than 2.5 percent of the oblast's land area is classified as forest (WB et al. 2005) and what remains is under serious threat.

In a related process of natural capital loss, the practice of burning manure for cooking and heating diverts it from its traditional use as fertilizer, further depleting soil fertility (MSDSP 2004). In the winter months, 96 percent of households in Gorno-Badakhshan burn manure and over half of all households burn it in the summers (MSDSP 2004, 6). Evidence of soil degradation is reflected, for example, in the 25 percent drop in potato yields in Gorno-Badakhshan from 2002 to 2003 (MSDSP 2004, 7).

Gorno-Badakhshan and Water   More than 90 percent of rural households in Gorno-Badakhshan do not enjoy the health benefits associated with having piped water in the home (WB et al. 2005), and nearly 60 percent of households lack access to piped water on their land or in their neighborhood (MSDSP 2004, 6).

Table 10.5
Average area of irrigated arable land per household in Gorno-Badakhshan (GBAO) by *raion*, 2004

Raion	Average area per household (hectares)
Darvoz	0.16
Ishkashim	0.44
Khorugh	0.03
Roshtkala	0.26
Rushon	0.39
Shughnon	0.31
Vanj	0.46
Total	0.30

Source: MSDSP 2004, 23.

Moreover, neither rain-fed nor irrigated farming offers livelihood security to farmers in Gorno-Badakhshan (table 10.5). The only location where some rain-fed farming is practiced is in the lowest-lying Darvoz Raion. Even there, however, rain-fed wheat yielded only one-third the harvests of irrigated wheat (MSDSP 2004, 30), a clear indication of the paucity of rain and need for irrigation. Unfortunately, the irrigation infrastructure in Gorno-Badakhshan is so limited that households in all *raiony* lack the irrigated arable land necessary to ensure food security.[12]

Like the lack of electricity and natural gas infrastructure, the lack of water piping and irrigation infrastructure is primarily the responsibility of the state, a form of neglect tantamount to environmental injustice.

Gorno-Badakhshan and Livestock   The wealthiest households in Gorno-Badakhshan own, on average, fewer than five sheep and four goats (MSDSP 2004, 24). In the case of sheep, thirty animals are required per person to ensure a sustainable livelihood (Breu and Hurni 2003, 43). It has become increasingly challenging to provide fodder for animals through the long winters (Ludi 2003, 23), which limits the maximum size of herds. Moreover, summer pastures are largely inaccessible due to the loss, during the Soviet period, of traditional knowledge associated with using camels and yaks for transport, combined with the present deterioration of tractors, neglected transportation infrastructure, and high fuel costs (MSDSP 2004, 22).

Table 10.6
Percentage of households with livestock in Gorno-Badakhshan (GBAO) that vaccinate, by livestock type and disease

		Cattle	Goats	Sheep	Chickens
Number of households owning animals		501	476	419	297
Percentage of households having their animals vaccinated	Anthrax	21	3	5	—
	Foot and mouth	58	8	14	—
	Emkar	12	1	3	—
	Brucellosis	15	4	9	—
	Newcastle	—	—	—	29
	Don't know	9	2	5	1

Source: MSDSP 2004, 25.

To these constraints must be added the lack of adequate state veterinary services, as indicated by the low percentages of households with livestock who vaccinate their animals. Due to the concentration of herds and flocks around villages, as noted above, animals are at a greater risk of becoming diseased and dying, all the more so for not having been vaccinated.[13] (See table 10.6.)

Cotton-Farming Households

Being among the most impoverished in the country, cotton-farming households generally suffer from an attendant lack of access to land, water, and livestock. Starting in Soviet times, the state placed great demands on farmers to produce Tajikistan's most lucrative export, "white gold" (WB 2005, 19). And yet this has not translated into livelihood security for cotton farmers. On the contrary, "Cotton farmers are poorer than non–cotton farmers despite increased [production] and higher international [cotton] prices in 2003 as compared to 1999" (WB 2005, 18). Cotton is grown in Sugd and Khatlon Oblasty and in the Raiony of Republican Subordination. Khatlon Oblast produces the most cotton and is also home to the greatest share of Tajikistan's impoverished households, who constitute 40 percent of the total population (UNDP 2003 in WB 2005). The overall poverty rate in Khatlon Oblast is 78 percent, second only to Gorno-Badakhshan (UNDP 2003 in WB 2005).

Moreover, because of the unjust nature of the cotton sector in Tajikistan, cotton-farming households deserve special consideration independent of being poor. Cotton magnates profit handsomely from the cotton sector, whereas the farmers themselves lose out because of inefficient machinery, poor storage facilities, and a monopsonistic system closely connected with the government. Farmers have only one buyer of cotton at the *raion* level, resulting in artificially low returns. Futures companies, which provide inputs such as seeds, fertilizers, and tools in exchange for compensation at harvest time, are driving thousands of cotton-farming households into debt by overvaluing their inputs and undervaluing the cotton output. As of 2005, total cotton debt was estimated to have reached \$220 million (Oxfam 2006, 4). Lacking access to credit and financial services and competitive sources of inputs, cotton-farming households are unable to extricate themselves from poverty.

Cotton Farmers and Land   Optimal agricultural lands, defined as those below 1,800 meters (5,900 feet) and with a slope of less than 10 percent, are dominated by cotton cultivation and located in Khatlon and Sugd Oblasty, as well as parts of the Raiony of Republican Subordination (see figure 10.10).[14] Indeed, 40 percent of arable lands are dedicated to cotton production (ICG 2005, 6).

Although the current government ostensibly completed privatization of land in January 2006, no significant changes were forthcoming in cotton-growing areas (MSDSP 2005, 9; Porteous 2003). Unfortunately, the term "privatization" is a misnomer: all land remains the property of the state, on long-term, inheritable lease to farmers; households may not sell their land or even use it as collateral (MSDSP 2005, 9). Farmers are rarely given land use certificates; in cotton-growing areas, they still have to belong to a farming association, which largely dictates what they plant, when they plant it, and to whom they sell it. If the state deems that their land is not being used properly, it can repossess it at any time (ICG 2005).

In a recent survey of rural households in Khatlon Oblast, 42 percent of respondents stated that they could not afford the cost of purchasing a land use certificate, which, including bribes, amounted to three months' salary (MSDSP 2005, 12–13). Indeed, some 90 percent of rural households have access only to home gardens or to "presidential lands" (Porteous 2003, 4–5).[15] These small plots, which average 0.23 hectare (0.57 acre) per household (Porteous 2003, 4–5), are not large enough to

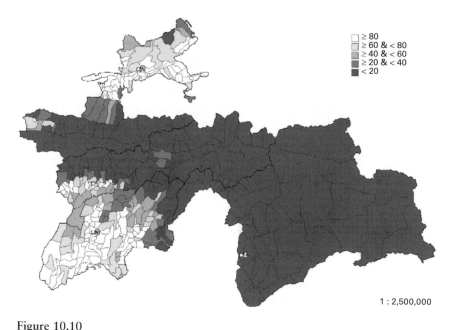

Figure 10.10
Map of percentage cover of optimal agricultural land in Tajikistan (WB et al. 2005).

ensure the food security of even one individual, for which, as noted above, the recognized threshold is 0.50 hectare (1.24 acres; Scherr 1999).[16]

Historical and current deforestation caused by cultivating crops on once-forested lands (extensification) has resulted in soil erosion and loss of soil fertility. Approximately 12 percent of the land currently under cultivation was once forested with *tugai*, pistachio, almond, and other broadleaf trees before the 1930s, when the Soviets began extensification (UNECE 2004, 127). In addition to the loss of biodiversity and forest-based livelihood strategies, the exposed soil has lost fertility for lack of the nitrogen fixing services of removed trees. Though deforestation has created over 88,000 hectares (217,400 acres) of cropland, much of it is eroded and unproductive, located on steep slopes (UNECE 2004, 138).

Erosion and salinization further compromise a high percentage of Tajikistan's arable land. Of a total of 739,100 hectares (1.83 million acres) of arable land in 2002, over 82 percent was eroded and 18 percent salinized (TSSC and ADB 2002 in UNECE 2004, 138). Furthermore, the United Nations Development Program reports that 50 million metric tons of arable soil continues to be lost annually to water and wind erosion and salinization (UNDP 2005, 43). In the 1980s, under a fully operational Soviet system that included clover-cotton and wheat-cotton crop rotations, 1 million metric tons of cotton was produced annually. Since then, the deteriorating irrigation infrastructure, loss of extension services, drought, erosion, and salinization have impacted soil fertility to such an extent that less than half that amount of cotton (453,000 metric tons) was produced in 2001 (ICG 2005). Indeed, a government specialist in Dushanbe admitted, under guarantee of anonymity, that the lack of crop rotation is "killing the land, and the Ministry of Agriculture is doing nothing" (ICG 2005).

Cotton Farmers and Water   In addition to domestic water insecurity, farming households lack the irrigation infrastructure necessary to ensure water security for their crops. In a recent survey conducted in Khatlon Oblast, only 13 percent of respondents had access to state irrigation systems (MSDSP 2005, 63).[17] Given the paucity of rain-fed lands in Khatlon (UNDP 2005), this represents a dire situation.

Cotton is a water-intensive crop, and the massive irrigation infrastructure built to maintain it extends to approximately 93 percent of Tajikistan's cultivated lands (FAO 2004). At present, great quantities of water are wasted through leaching and evaporation from canals. Indeed, approximately 96 percent of Tajikistan's irrigation network consists of open cement or dirt-furrow canals that often lack proper drainage systems (FAO 2004). Some fields remain parched, while others become waterlogged, resulting in the salinization described above.

Cotton Farmers and Livestock   Farmers in Khatlon Oblast often have to seek out private veterinary services to supplement inadequate state services. Though the state is supposed to provide free vaccinations for all households' livestock, these are reportedly offered only in the case of anthrax and only *after* the outbreak of epidemics (MSDSP 2005, 45). Moreover, those farmers vaccinating their livestock through state veterinarians can only afford to vaccinate them against one of the three major diseases: anthrax, brucellosis, and foot-and-mouth disease (MSDSP 2005, 46), leaving their herds vulnerable to disease and death. (See figure 10.11.)

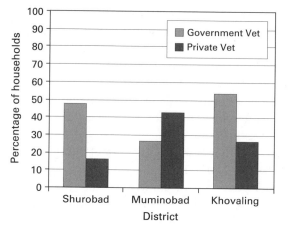

Figure 10.11
Percentage of households utilizing livestock vaccinations in Khatlon Oblast, Tajikistan, by district (MSDSP 2005, 46).

In sum, impoverished households in general are most likely to be the victims of environmental injustice in Tajikistan. Moreover, these households were disaggregated into those headed by single women, those living in Gorno-Badakhshan, and those engaged in cotton farming, for whom explanations of environmental injustice exist in addition to general indicators of poverty. Although it is entirely plausible that ethnic minorities are poorer than ethnic Tajiks, and that they, as a result, may suffer disproportionately from environmental injustice, studies have not yet disaggregated wealth based on ethnic identity; nor has ethnicity been studied as an independent factor concerning environmental injustice. The national government was shown to be responsible for much of the environmental injustice perpetrated on rural households, largely through lack of governance, a theme further explored in the next section.

Lack of National-Level Environmental Governance

At present, the highly centralized and authoritarian Tajik national government lacks effective processes of environmental governance. As described above, the government perpetuates some forms of environmental injustice to the detriment of the rural population. To its credit, the government has enacted significant environmentally related laws (for a comprehensive list, see UNECE 2004, 179–184), but these have not been effectively implemented or enforced.

If fully implemented, the State Environment Program, which "calls for a balance to be struck between economic activity and the carrying capacity of the environment," while "preventing land erosion; [conducting] reforestation; expanding specially protected territories; restoring good quality of air, water, and other resources [and] introducing energy saving technologies into industry" (UNECE 2004, 28, 29), and the Poverty Reduction Strategy Paper (PRSP; Tajik Government 2002), which focuses on the creation of new jobs and recognizes that "natural disasters, water pollution, soil erosion, and desertification have a serious impact on the poor" (UNECE 2004, 18), would help protect natural capital and support rural livelihoods. The PRSP most concretely addresses environmental concerns through agricultural sector reform. Although the paper formed the basis of Tajikistan's 2004–2006 planned public investments, the government only allocated $125 million of the $690 million required by its own budget (UNECE 2004, 18). Neither of these national initiatives, since being approved in 2002, has succeeded in securing and guaranteeing access to and control of natural capital for rural households.

Ratification of international environmental conventions has been equally ineffective in ensuring environmental justice. In 1997, Tajikistan ratified the UN Convention on Biological Diversity and the UN Convention to Combat Desertification and, in 1998, the UN Framework Convention on Climate Change (UN 1992a; UN 1994; UN 1992b). After ratification, however, it took six years for national action plans to be written. In 2005, an assessment report highlighted the need for national capacity building in environmental governance and emphasized the seriousness of environmental degradation; the lack of environmental experts; the paucity of high-quality environmental and technical education; the lack of accurate and accessible environmental information; and the ineffectiveness of current legislation (UNDP 2005). Clearly, Tajikistan has many obstacles to surmount to ensure environmental justice, sustainable livelihoods, and peace for its population.

It is significant that, in 2001, Tajikistan also ratified the Aarhus Convention on Access to Information, Public Participation in Decision Making, and Justice in Environmental Matters, which represents an essential component of effective environmental governance (UNECE 1998). The Organization for Security and Cooperation in Europe (OSCE) conducted two regional workshops on this convention in 2004, stressing public involvement in what have traditionally been governmental decision-making processes (OSCE 2006). Unfortunately, that same year, the UN Economic Commission for Europe reported that "the general public, including NGOs, does not currently have any role in the legislative process, except for the extremely rare occasions when one of the State bodies may decide at its own initiative to seek public opinion on the draft [legislation]" (UNECE 2004, 31).

Though limited initiatives have been undertaken, environmental governance structures and processes have thus far failed to protect and manage natural capital for the benefit of rural households in Tajikistan. Despite the efforts of international organizations and, to a limited extent, local civil society in promoting environmental governance, some of the national government's policies have resulted in further depleting natural capital, degrading the environment, and perpetuating environmental injustice.

Environmental Justice Movement in Tajikistan: The Beginnings of a Dialogue?

Is the concept of an environmental justice movement relevant to Tajikistan at this time? Despite being a signatory to the Aarhus Convention and having enacted extensive environmental legislation, Tajikistan remains a highly authoritarian state in which the average citizen's access to environmental information and participation in environmental governance processes are severely limited. In order for meaningful, participatory dialogue on environmental justice, sustainable livelihoods, and peace to emerge in Tajikistan, the national government must change its ways—as indeed must international organizations and local civil society.

Role of International Organizations
International organizations that are active in Tajikistan, such as the Agency for Technical Cooperation and Development (ACTED), the Aga Khan Foundation, FAO, Agency for Technical Cooperation (GTZ), Mercy Corps, the Open Society Institute, OSCE, Oxfam, and UNDP, represent a combination of sustainability values and desperately needed funding for both the national government and NGOs (Abdusalyamova 2002, 1). The interplay between these international, national, and local stakeholders can stimulate dialogue and action concerning concepts like

environmental, social, and economic justice, even if such dialogue and action are at first partly motivated by the need for funding.

As an international framework of shared principles for sustainable development, and the product of the most inclusive and participatory consultative process ever undertaken for an international document (including consultations in Central Asia and Tajikistan in the 1990s), the Earth Charter is well situated to provide the common ground necessary for the dialogue on sustainability values and action mentioned above. Indeed, the processes of designing, implementing, and evaluating sustainable development projects in Tajikistan involve a delicate exchange of values between international, national, and local organizations. Such dialogue has the potential to result in the open exchange of ideas, mutual learning, and locally adapted sustainable development. In the context of eradicating poverty, the Earth Charter explicitly recognizes the need to build capacity for sustainable livelihoods, the right to access natural capital, and the attendant responsibility to manage natural resources sustainably (ECC 2000, 9b, 9a, 5e), all good starting points for ensuring environmental justice.

That said, international organizations should be careful not to overextend themselves into Tajikistan's development discourse, policies, programs, and projects. In 2006, there were at least 110 sustainable development projects under way in Tajikistan, totaling $591 million in investments (Putnam and Mukhamadiev 2006), nearly one-third of the country's GDP (UNRC 2006). Of these monies, some 25 percent went to fund energy projects and 20 percent each to fund agriculture, water, and health projects (extrapolated from UNRC 2006). Although such projects are essential for Tajikistan's development, international organizations must refrain from imposing their own agendas and establishing themselves as indispensable environmental governance institutions. Instead, they should work toward empowering national and local stakeholders to such an extent that international support is seldom required, if at all.

Role of Local Civil Society

Although there are some 600 NGOs in Tajikistan (Abdulsalyamova 2002, 1), few of these are environmental NGOs (Putnam and Mukhamadiev 2006), and their activities generally lack long-term, coordinated planning (Farmer and Farmer 2001). Indeed, even though implementation of over 40 percent of the above-mentioned sustainable development projects involved Tajik organizations, only five local NGOs played a role. Furthermore, NGOs are concentrated in urban areas, with some of them aligning their activities with those of the government (Farmer and Farmer 2001). Others are established primarily for the financial benefit of their staffs, tailoring their activities to available grants (Abdulsalyamova 2002, 2). Indeed, such practices disassociate some NGOs from local, rural priorities, leading citizens of Tajikistan to distrust them. On the other hand, NGOs such as the Foundation to Support Civil Initiatives are focused on the "development and realization of programs for sustainable development and environmental sustainability at the local level" (Burkhanova 2005, 161), furthering the values underpinning environmental justice.

Civil society in Tajikistan must strive for integrity, autonomy, and coordinated empowerment to maintain the trust of the people and engage in meaningful sustainable development activities. Currently, it is still highly dependent on international sources of funding and is not empowered to participate in national environmental governance processes. In this regard, international organizations must refrain from fostering financial dependence, just as the national government must give greater support and autonomy to civil society actors. With well-informed and empowered civil society participation, meaningful dialogue on sustainable development and environmental justice in Tajikistan can begin.

Key Questions for an Environmental Justice Movement in Tajikistan

Various environmentalisms evolve out of diverse situations, and that a single definition of the [environmental justice] movement, its motivations, or its aims [should emerge] is not only impossible, but unnecessary and counterproductive.
—David Schlosberg, *Environmental Justice and the New Pluralism*, 1999

If it is to be meaningful and durable, a contextualized definition of environmental justice in Tajikistan must emerge through inclusive, participatory, and well-informed dialogue. Such dialogue itself would constitute the beginnings of an environmental justice movement. As stakeholders discuss and act upon the concept of environmental justice, they will, at a minimum, have to consider three sets of questions:

• How is the environment defined? What is the relationship between humans and the environment? What aspects of the environment do local

citizens value? Are individuals empowered to access, sustainably manage, and protect their own environment?
• How is justice defined? What are people's expectations of their government and courts in terms of legislating and enforcing just laws? To what extent do the government and judicial system fulfill these expectations? Is the government democratic, transparent, and accountable?
• What is the state of civil society? Are local women and men able and willing to take the risk of advocating for change? Are they empowered with reliable information and access to participate in important decision-making processes? Will decision makers hear their grievances and proposals?

I dedicate this chapter to rural Central Asian families who struggle to stay together in the face of great adversity. May it and this book stimulate meaningful dialogue on environmental justice, sustainable livelihoods, and peace not only in Tajikistan and Central Asia, but throughout the former Soviet Union as well.

Notes

1. Natural capital, a concept from the sustainable livelihoods discourse, is defined by the United Kingdom's Department for International Development as "the natural resource stocks from which resource flows and services useful for livelihoods are derived" (DFID 1999, 2.3.3).

2. Tajikistan is made up of its capital, Dushanbe, and four oblasty. Sugd (formerly Leninobod) Oblast is located in the northwest; the Raiony of Republican Subordination (RRS) surround and are administered by Dushanbe; Khatlon Oblast (formed from the joining of Khatlon and Kulob regions) lies in the southwest; and mountainous Gorno-Badakhshan Autonomous Oblast (GBAO) makes up eastern Tajikistan.

3. The research that informs this chapter can be found in my master's thesis, "Linking Natural Capital, Rural Livelihoods, and Conflict: Toward Governance for Environmental Security and Peace in Tajikistan" (Stucker 2006), http://www.untj.org/library/?mode=details&id=327.

4. Approximately 50,000 landslides occur each year in Gorno-Badakhshan; the oblast has identified 700 families who are in immediate need of being relocated because of them, in addition to more than 10,000 families that may become environmental migrants in the next five years (UNECE 2004, 17).

5. According to the State Hydro-Meteorological Agency of Tajikistan, various regions within the country received 80 percent of regular precipitation in 1999, 40–65 percent in 2000, and only 40–60 percent in 2001 (UNDP 2005, 43).

6. A recent survey of long-term glacier and climate variability in the mountain regions of the former Soviet Union indicates a general warming trend, with glacier sizes diminishing throughout the area (Solomina 2005, 70). Indeed, temperature increases and changes in precipitation have been measured in the Pamir Mountains of Gorno-Badakhshan and threaten the vast water reserves stored in the oblast's 8,492 glaciers. Between 1961 and 1990, the temperature increased by 0.5 degrees Celsius (0.9 degrees Fahrenheit) and the Fedshenko Glacier, at 70 kilometers (43.5 miles) one of the region's largest, lost 1 kilometer (1,100 yards) of its length (Breu and Hurni 2003, 9). Though not the focus of this chapter, climate change is a poignant form of environmental injustice, with vulnerable peoples often contributing the least to greenhouse gas emissions, yet being the most severely impacted and least able to adapt.

7. These statistics are based on unpublished background information for a 2006 livelihood project proposal by German Agro Action (WHH) in Tajikistan.

8. Rural households in Tajikistan generate, on average, 25 percent of their incomes from nonagricultural sources (MSDSP 2004, 13) so they should only require 75 percent of the minimum land requirements to ensure food security. Even considering this adjusted threshold, they are still left with less than one-fourth the necessary land.

9. The exchange rate of Tajik somoni to dollars has been relatively steady for the past few years at approximately 3:1.

10. Estimates of Tajikistan's poverty rates range from 68 percent (UNRC 2006) to 83 percent (Tajik Government 2002).

11. Gorno-Badakhshan, though sparsely populated, has experienced significant population growth. From 1926, just before the Soviets established control, to 2000, its population nearly quadrupled, from 56,000 to 220,000 (Breu and Hurni 2003, 10). The Soviets populated the oblast in an effort to solidify their territorial control; Gorno-Badakhshan experienced a spurt in population during and after the civil war, when it provided a relatively safe haven from the conflict.

12. As one might expect, the oblast's capital, Khorugh, and Darvoz Raion have the least amount of irrigated arable land per household second and third only to Murghob Raion, which, being on a high mountain plateau, has no irrigated land at all.

13. A common poultry disease in Tajikistan, Newcastle disease wiped out entire chicken flocks in Gorno-Badakhshan in 2003. Unfortunately, one year later, only 29 percent of the oblast's chickens had been vaccinated (MSDSP 2004, 25).

14. This "optimal" definition of agricultural land, however, does not include such critical factors as water availability and soil fertility, key forms of natural capital.

15. Presidential lands, small plots on the margins of state farms, were made available by decrees in 1995 and 1997 in attempts to stem food insecurity (Porteous 2003, 5).

16. Even if land requirements are reduced by 25 percent, the estimated portion of rural household income from nonagricultural sources, average household plots only amount to three-fifths of the adjusted threshold for food security for one individual.

17. Depending on the *raion*, between 0 and 8 percent of households surveyed had access to and used primary canals, while 0 to 23 percent used secondary canals.

References

Abdusalyamova, Lola. 2002. "NGOs in Central Asia." *Alliance* 7 (2): 23–25. www.untj.org/files/reports/NGOs%20in%20Central%20Asia.pdf/   (accessed April 5, 2007).

Azamova, Asal. 2001. "The Military is in Control of Drug Trafficking in Tajikistan." *Moscow News*, May 30.

Baran, Z., F. S. Starr, and S. E. Cornell. 2006. *Islamic Radicalism in Central Asia and the Caucasus: Implications for the EU, Silk Road Paper*. Washington, DC: Central Asia-Caucasus Institute (CACI) and Silk Road Studies Program (SRSP).

Breu, T., and H. Hurni. 2003. *The Tajik Pamirs: Challenges of Sustainable Development in an Isolated Mountain Region*. Berne: Center for Development and Environment (CDE), University of Berne. http://www.untj.org/files/reports/The%20Tajik%20Pamirs%20english-low%20res.pdf/ (accessed July 3, 2006).

Burkhanova, Muazama. 2005. "Environmental Problems and Sustainable Development in Tajikistan." In Peter Blaze Corcoran, Mirian Vilela, and Alide Roerink, eds., *The Earth Charter in Action*, 159–160. Amsterdam: KIT. http://earthcharterinaction.org/eci_book.shtml/ (accessed April 5, 2007).

Carius, A., M. Feil, and D. Tänzle. 2003. *Addressing Environmental Risks in Central Asia*. Berlin: United Nations Development Fund, Regional Bureau for Europe and the Commonwealth of Independent States (CIS). http://www.iisd.org/pdf/2003/envsec_undp_ca_study.pdf/ (accessed July 5, 2006).

Carney, Diana, M. Drinkwater, T. Rusinow, K. Neefjes, S. Wanmali, and N. Singh. 1999. *Livelihood Approaches Compared*. Department for International Development (DFID). http://www.livelihoods.org/cgi-bin/dbtcgi.exe/ (accessed May 6, 2006).

Chambers, Robert, and Gordon R. Conway. 1991. *Sustainable Rural Livelihoods: Practical Concepts for the 21st Century*. Discussion Paper no. 296. Institute of Development Studies (IDS). University of Sussex, Brighton, UK.

Cincotta, R., R. Engelman, and D. Anastasion. 2003. *The Security Demographic: Population and Civil Conflict after the Cold War*. Washington, DC: Population Action International (PAI). http://www.populationaction.org/securitydemographic/pdfs/SecurityDemographic.PDF/ (accessed July 20, 2006).

Cornell, Svante E., and Regine A. Spector. 2002. "Central Asia: More than Islamic Extremists." *Washington Quarterly* 25 (1): 193–206. http://www.twq .com/02winter/spector.pdf/ (accessed June 1, 2006).

de Martino, L., A. Carlsson, G. Rampolla, I. Kadyrzhanova, P. Svedberg, N. Denisov, V. Novikov, P. Rekacewicz, O. Simonett, J. Fernandez Skaalvik, D. del Pietro, D. Rizzolio, and M. Palosaari. 2005. Environment and Security Initiative in Central Asia (ENVSEC), *Environment and Security, Transforming Risks into Cooperation, Central Asia, Ferghana/Osh/Khujand area.* Geneva: UNEP, UNDP, OSCE, and NATO.

De Nuebourg, C., and C. Namazie. 1999. "Labor and Poverty in Tajikistan." Background paper prepared for World Bank, Washington, DC.

Department for International Development (DFID). 1999. *Sustainable Livelihoods Guidance Sheets.* London. http://www.livelihoods.org/info/info _guidancesheets.html/ (accessed June 2, 2006).

Earth Charter Commission (ECC). 2000, March. *Earth Charter.* Paris. http:// www.earthcharter.org/files/charter/charter.pdf/ (accessed May 5, 2006).

Ellis, Frank. 1998. "Household Strategies and Rural Livelihood Diversification." *Journal of Development Studies* 35 (1): 1–38.

Eurasia Insight. 2006. "Moscow Market Tragedy Refocuses Attention of Tajik Labor Migration Issue." February 28. http://www.eurasianet.org/departments/ civilsociety/articles/eav022806.shtml/ (accessed June 4, 2006).

European Commission Humanitarian Aid Office (ECHO). 2003. *National Nutrition, Water and Sanitation Survey.* Dushanbe.

Falkingham, Jane, and Asian Development Bank (ADB). 2000. *Country Briefing Paper: Women and Gender Relations in Tajikistan.* Dushanbe: Asian Development Bank.

Farmer, Andrew M., and Alma A. Farmer. 2001. "Developing Sustainability: Environmental Non-governmental Organizations in Former Soviet Central Asia." *Sustainable Development* 9 (3): 136–148. http://www3.interscience.wiley.com/ cgi-bin/abstract/85007467/ABSTRACT/ (accessed April 5, 2007).

Foreman, Christopher. 1998. *The Promise and Peril of Environmental Justice.* Washington, DC: Brookings Institution Press.

Food and Agriculture Organization (FAO). 2004. *AquaStat, Tajikistan.* http:// www.fao.org/ag/agl/aglw/aquastat/countries/tajikistan/print1.stm/ (accessed February 14, 2006).

Freedom House. 2007. *Countries at the Crossroads, Tajikistan Report.* http://www.freedomhouse.org/template.cfm?page=140&edition=8&ccrpage=37 &ccrcountry=170/ (accessed March 8, 2008).

Gerstle, Daniel. 2004. "Fighting Hunger and Heroin on the Afghan-Tajik Border." *Eurasia Insight,* November 11. http://www.eurasianet.org/departments/ insight/articles/eav111104_pr.shtml/ (accessed January 17, 2006).

Gretsky, Sergei. 1995. "Civil War in Tajikistan: Causes, Developments, and Prospects for Peace." In Roald Z. Sagdeev and Susan Eisenhower, eds., *Central*

Asia: Conflict, Resolution, and Change, 218–248. Chevy Chase, MD: Center for Political and Strategic Studies (CPSS) Press. http://www.eisenhowerinstitute.org/ programs/globalpartnerships/securityandterrorism/coalition/regionalrelations/ ConflictBook/Gretsky.htm/ (accessed May 31, 2006).

Holifield, Ryan. 2001. "Defining Environmental Justice and Environmental Racism." *Urban Geography* 22 (1): 78–90. http://www.bellpub.com/ug/2001/ ad010105.pdf/ (accessed April 5, 2007).

International Crisis Group (ICG). 2001. *Incubators of Conflict: Central Asia's Localized Poverty and Social Unrest.* Asia Report no. 16. http://www.crisisgroup .org/ (accessed July 29, 2006).

International Crisis Group (ICG). 2003. *Tajikistan: A Roadmap for Development.* Asia Report no. 51. Osh, Kyrgyzstan. http://www.crisisgroup.org/home/ index.cfm?id=1447&l=1/ (accessed February 15, 2006).

International Crisis Group (ICG). 2005. *The Curse of Cotton: Central Asia's Destructive Monoculture.* Asia Report no. 93. Bishkek, Kyrgyzstan. http://www .crisisgroup.org/home/index.cfm?id=3294&l=1/ (accessed February 14, 2006).

International Organization of Migration (IOM). 2001. "IOM Study Reveals Trends in Trafficking in Women from Tajikistan." August 17. Dushanbe. http:// www.iom.int/tajikistan/news170801.htm (accessed February 18, 2006).

Khudonazar, Davlat. 1995. "The Conflict in Tajikistan: Questions of Regionalism." In Roald Z. Sagdeev and Susan Eisenhower, eds., *Central Asia: Conflict, Resolution, and Change*, 249–263. Chevy Chase, MD: Center for Political and Strategic Studies (CPSS) Press. http://www.eisenhowerinstitute.org/programs/ globalpartnerships/securityandterrorism/coalition/regionalrelations/ConflictBook/ Khudonazar.htm/ (accessed May 31, 2006).

Lillis, Joanna. 2006. "Tajikistan: No Surprises in Presidential Election." *Eurasia Insight*, November 6. http://www.eurasianet.org/departments/insight/articles/ eav110606a.shtml/ (accessed January 15, 2007).

Ludi, Eva. 2003. "An Economy in Transition: Managing High Pastures in the Eastern Pamirs." In T. Breu and H. Hurni, eds., *The Tajik Pamirs: Challenges of Sustainable Development in an Isolated Mountain Region*, 22–23. Berne: Center for Development and Environment (CDE), University of Berne. http://www .untj.org/files/reports/The%20Tajik%20Pamirs%20english-low%20res.pdf/ (accessed July 3, 2006).

Makarenko, Tamara. 2000. "Crime and Terrorism in Central Asia." *Jane's Intelligence Review* 12 (7): 16–17.

McGlinchey, Eric. 2005. "The Making of Militants: The State and Islam in Central Asia." *Comparative Studies of South Asia, Africa and the Middle East* 25 (3): 554–566. http://muse.jhu.edu/journals/comparative_studies_of_south_asia_africa _and_the_middle_east/v025/25.3mcglinchey.pdf/ (accessed May 31, 2006).

Morgounov, A. and Zuidema, L. 2001. *The Legacy of the Soviet Agricultural Research System for the Republics of Central Asia and the Caucasus.* Research Report no. 20. The Hague: International Service for National Agricultural

Research. ftp://ftp.cgiar.org/isnar/Publicat/PDF/rr-20.pdf/ (accessed February 14, 2006).

Mountain Societies Development Support Program (MSDSP). 2004. *2003 Baseline Survey of Gorno-Badakhshan Autonomous Oblast, Tajikistan.* Dushanbe: Aga Khan Foundation (AKF). http://www.untj.org/files/reports/ GBAO%20baseline%20survey%202003.pdf/ (accessed August 30, 2006).

Mountain Societies Development Support Program (MSDSP). 2005. *Client Perception Survey 2005: Baseline Findings Khatlon Oblast, Tajikistan.* Dushanbe: Aga Khan Foundation (AKF). http://www.untj.org/files/reports/Khatlon %20Client%20Perception%20Survey%202005.pdf/ (accessed August 30, 2006).

National Intelligence Council (NIC). 2004. *Mapping the Global Future: Report of the National Intelligence Council's 2020 Project.* Washington DC: Central Intelligence Agency. http://www.foia.cia.gov/2020/2020.pdf/ (accessed June 2, 2006).

Neefjes, Koos. 2000. *Environments and Livelihoods: Strategies for Sustainability.* Oxford: Oxfam.

Olimova, Saodat. 2004. Regionalism and Its Perception by Major Political and Social Powers of Tajikistan. In Luigi De Martino, ed., *Tajikistan at a Crossroad: The Politics of Decentralization,* 85–118. Situation Report no. 4. Geneva: CIMERA.

Organization for Security and Cooperation in Europe (OSCE). 2006a. "Environmental Activities: Improving Participation in Environmental Matters." http:// www.osce.org/tajikistan/13491.html/ (accessed June 17, 2006).

Organization for Security and Cooperation in Europe (OSCE). 2006b. *Republic of Tajikistan: Parliamentary Elections of 27 February and 13 March 2005:* OSCE/ODIHR [Office for Democratic Institutions and Human Rights] Election Observation Mission Final Report. Warsaw: OSCE and ODIHR. http://www .osce.org/documents/odihr/2005/05/14852_en.pdf/ (accessed January 15, 2007).

Oxfam. 2006. *Community Situation Indicators: Identifying Priorities: School Desk or Cotton Field?* Dushanbe. http://www.untj.org/files/reports/CSI_bulletin-Children&Cotton_eng.pdf/ (accessed June 10, 2006).

Porteous, Obie. 2003. *Land Reform in Tajikistan: From the Capital to the Cotton Fields.* Dushanbe: Action Against Hunger (AAH). http://www.untj.org/files/ reports/report_land_reform_aah_ec.pdf/ (accessed August 31, 2006).

Putnam, Evelynn, and Bakhtiyor Mukhamadiev. 2006. "ESTH Projects in Tajikistan." *Central Asian Environmental, Science, Technology and Health (ESTH) Newsletter,* February. Almaty, Kazakhstan.

Rashid, Ahmed. 2000. "Islamic Movement of Uzbekistan's Incursion Assists the Taliban." *Central Asia-Caucasus Analyst,* September 13. http://www.cacianalyst .org/view_article.php?articleid=132/ (accessed May 31, 2006).

Rashid, Ahmed. 1994. *The Resurgence of Central Asia: Islam or Nationalism?* London: Zed Books.

Scherr, Sara. 1999. *Soil Degradation: A Threat to Developing Country Food Security?* Washington DC: International Food Policy Research Institute (IFPRI).

Schlosberg, David. 1999. *Environmental Justice and the New Pluralism: The Challenge of Difference for Environmentalism.* Oxford: Oxford University Press.

Solomina, Olga N. 2005. "Glacier and Climate Variability in the Mountains of the Former Soviet Union during the Last 1,000 Years." In Uli M. Huber, Harald K. M. Bugmann, and Mel A. Reasoner, eds., *Global Change and Mountain Regions: An Overview of Current Knowledge*, 61–72. Dordrecht, Netherlands: Springer.

Stucker, Dominic. 2006. "Linking Natural Capital, Rural Livelihoods, and Conflict: Toward Governance for Environmental Security and Peace in Tajikistan." Master's thesis, University for Peace, Costa Rica. http://www.untj.org/library/?mode=details&id=327/ (accessed April 22, 2007).

Tajik Government. 2002. *Poverty Reduction Strategy Paper.* Dushanbe: International Monetary Fund. http://www.imf.org/External/NP/prsp/2002/tjk/01/061902.pdf#search=%22Tajikistan%20Poverty%20Reduction%20Strategy%20Paper%22/ (accessed July 20, 2006).

Tajik State Statistical Committee (TSSC) and Asian Development Bank (ADB). 2002. *Poverty Reduction Monitoring Survey.* Dushanbe: Asian Development Bank.

United Nations (UN). 1992a. *UN Convention on Biological Diversity (UNCBD).* New York. http://www.cbd.int/convention/convention.shtml/ (accessed February 2, 2006).

United Nations (UN). 1992b. UN Framework Convention on Climate Change (UNFCCC). New York. http://unfccc.int/essential_background/convention/background/items/2853.php/ (accessed February 2, 2006).

United Nations (UN). 1994. *UN Convention to Combat Desertification (UNCCD).* Paris. http://www.unccd.int/convention/text/pdf/conv-eng.pdf/ (accessed February 2, 2006).

United Nations (UN). 2004. Department of Peacekeeping Operations, Cartographic Division, *Political Map of Tajikistan.* New York.

United Nations Development Program (UNDP). 2003. *Tajikistan Living Standards Survey.* Dushanbe.

United Nations Development Program (UNDP). 2005. *Report and Action Plan on Building National Capacity to Implement Commitments of the Republic of Tajikistan on Global Environmental Conventions.* Dushanbe. http://www.undp.tj/Publications/NCSA%20Action%20Plan%20in%20English%20final%20with%20design.pdf/ (accessed February 10, 2006).

United Nations Economic Commission for Europe (UNECE). 1998. *"Aarhus Convention,"* UNECE Convention on Access to Information, Public Participation in Decision-making and Access to Justice in Environmental Matters. Aarhus, Denmark. http://www.unece.org/env/pp/ (accessed February 1, 2006).

United Nations Economic Commission for Europe (UNECE). 2004. Committee on Environmental Policy, *Environmental Performance Reviews: Tajikistan*. New York: United Nations.

United Nations Office for the Coordination of Humanitarian Affairs (UNO-CHA). 2006. *Tajikistan: Year in Review 2005*. http://www.irinnews.org/report .asp?ReportID=51070&SelectRegion=Asia/ (accessed February 18, 2006).

United Nations Resident Coordinator, Tajikistan (UNRC). 2006. *Moving Mountains: The UN Appeal for Tajikistan, 2006*. Dushanbe.

United Nations World Water Assessment Program (UNWWAP). 2003. *UN World Water Development Report: Water for People, Water for Life*. Paris: United Nations Educational, Scientific and Cultural Organization (UNESCO) and Berghahn Books. http://www.unesco.org/water/wwap/wwdr/index.shtml/ (accessed November 19, 2005).

World Bank (WB). 2000. *Tajikistan Gender Profile*. Washington, DC. http:// siteresources.worldbank.org/INTECAREGTOPGENDER/Countries/20577404/ TajikGenderProfile.pdf/ (accessed May 20, 2006).

World Bank (WB). 2005. *Republic of Tajikistan Poverty Assessment Update*. Washington, DC. http://www.untj.org/files/reports/Tajikistan%20Poverty %20Assessment%20Update.pdf/ (accessed May 20, 2006).

World Bank (WB), University of Southampton, UK, Tajik State Statistical Committee (SSC), and Department for International Development (DFID). 2005. *Socio-Economic Atlas of Tajikistan*. Washington, DC: World Bank. http:// siteresources.worldbank.org/INTTAJIKISTAN/Resources/atlas_11.pdf/ (accessed May 20, 2006).

World Health Organization (WHO). 2006a. *Health for All Database*. http:// data.euro.who.int/hfadb/ (accessed August 5, 2006).

World Health Organization (WHO). 2006b. *Working Together for Health—The World Health Report 2006*. Geneva: WHO Press. http://www.who.int/whr/2006/ en/ (accessed September 5, 2006).

Zibung, Daniel. 2003. "The Natural Environment and Its Potential Energy: A Precondition for Development." In T. Breu and H. Hurni, eds., *The Tajik Pamirs: Challenges of Sustainable Development in an Isolated Mountain Region*, 35–36. Berne: Center for Development and Environment (CDE), University of Berne. http://www.untj.org/files/reports/The%20Tajik%20Pamirs %20english-low%20res.pdf/ (July 3, 2006).

Conclusion

Julian Agyeman and Yelena Ogneva-Himmelberger

Our intention in embarking on this project was to answer two related overarching questions and, in so doing, to stimulate further research:

1. To what extent are increased popular environmental awareness and associated activism driving public policy and planning in the former Soviet republics?
2. Are there emergent, separate brown (environmental justice) and green (environmentally sustainable development) agendas or are these joining together in a single just sustainability or human security agenda?

Have we answered these questions? Yes, we have, but with the caveats set out in the introduction to this volume, namely, that, given the vastness of the former Soviet Union with its immense sociocultural, political, environmental, and ethnic diversities, we can make no claim to be comprehensive in terms of geographical spread, analysis, or representation of environmental justice, (environmentally) sustainable development, or just sustainability and human security in the former Soviet Union. Instead, what we have sought to do is to begin a conversation and, we hope, also a research agenda on the growing global awareness of environmental justice, sustainable development, just sustainability, and human security,—and what shape, focus, and trajectory resultant activism and public policy and planning are taking, or might take, within the republics of the former Soviet Union.

Key Points

First, as Brian Donahoe shows, the biggest of the FSU republics, Russia, has progressive legal frameworks for environmental protection and indigenous people's rights. However, they have been rendered ineffective by a relentless recentralization of power, by the state's unwillingness to implement laws, and by a systematic evisceration of laws on environmental protection and indigenous rights through sweeping changes buried in omnibus bills. Nowhere is this more evident than in issues of control over land and natural resources, as many of our contributions demonstrate.

Second, and related, as Henry shows, Russian civil society actors embraced sustainable development in the early 1990s, but with the federal government's greater centralization of power under Putin-Medvedev, and their prioritization of economic development, most Russian environmental actors and organizations focused on green issues, conservation, or *environmental* sustainability, as opposed to just sustainability or human security. This has limited the potential and scope of sustainable development to the possible exclusion of those more interested in the social justice aspects, discouraging public participation and making concerted government action far less likely. On the other hand, as shown in chapters 6, 7, and 9, the environmental movements in Latvia, Kazakhstan (Berezovka), and Estonia, with their strong political, notably antitotalitarian and antibureaucratic character, have made the fight for a clean environment a fight for human rights. There, at least, environment and politics are intertwined.

Third, emerging and different sustainability discourses on Sakhalin Island and elsewhere, are configured around different actors (indigenous, local, and transnational) and different networks, which are creating a new internationalism and a call for environmental justice. As Jessica Graybill (chapter 3) explains:

Concerns are being raised by different sets of actors with different interests in— and different visions for–creating "sustainable Sakhalin." Multiple actors with different visions of Sakhalin's future are shaping its socioeconomic, cultural, and environmental landscapes. Thus a suite of roles and practices of engagement with issues of sustainability is developing among different communities on the island. This is reflected in the emergence of multiple new sociocultural and environmental movements on post-Soviet Sakhalin. Although often discussed in terms typical of the larger environmental justice movement (e.g., environmental degradation, environmental racism, environmental justice as social justice; see Benford 2005; Pellow and Brulle 2005), Sakhalin's environmental justice movements do not always overlap and sometimes contest central issues among themselves.

Not all of our authors agree, however, that there is a stand-alone environmental justice movement in any FSU republic. Maaris Raudsepp,

Mati Heidmets, and Jüri Kruusvall (chapter 9), for example, argue that environmental justice is embodied in the larger context of economic and social policies of the Estonian National Strategy on Sustainable Development (SE21). Susan Crate (chapter 8) argues that the recipe for the success of an environmental justice movement is in the combination of the following ingredients—a strong urban base, international contacts, local leadership, and the knowledge and the ability to take advantage of the existing legislation detailed by Donahoe (chapter 1) and mentioned above in the first key point. Crate concludes, however, that an environmental justice framework does not exist in Russia because some of these central ingredients are currently lacking. Katherine Metzo (chapter 5) and Kate Watters (chapter 8) emphasize the critical role of local leadership in particular.

Fourth, as noted in chapters 3, 6, and 9, culture, nationalism, environmentalism, and justice play complex, interacting roles in Russia (see third key point), Latvia (the "singing revolution"), and Estonia ("viability of the Estonian cultural space" is a principal theme in the Estonian National Strategy on Sustainable Development), three very different FSU republics.

Fifth, many of our authors (Graybill, Metzo, Watters, Crate, Steger, Raudsepp, Heidmets, Kruusvall, and Stucker in chapters 3 and 5–10) noted that there was evidence of what we called a "justice-informed discourse." Most conceded, however, that this was not a grassroots, "bottom-up" discourse, coming from within communities; it was more likely an imported discourse, fostered by international NGOs such as Global Response in the Tunka National Park, World Wildlife Fund and Pacific Environment in Sakhalin, Crude Accountability in Kazakhstan (Berezovka), and Mercy Corps and Oxfam in Tajikistan. These organizations play a role not only in discourse formation but also in funding and training on Sakhalin Island and in Tunka, Latvia, Kazakhstan, Estonia, and Tajikistan. Raudsepp, Heidmets, and Kruusvall (chapter 9) talk of the Estonian civil society groups "'shadowing' Western groups" to help them "gain access to Western informational and material resources."

Sixth, industrial development is seen as the main source of environmental injustice. This is most often gas or oil, as in Sakhalin, Tunka, Kazakhstan, and Azerbaijan, or diamonds, as in the Viliui Sakha regions. What is interesting here is how most of our authors implicitly point to the global-local connection behind environmental injustice. Shannon O'Lear (chapter 4) does so explicitly: "Environmental injustice is experienced locally, but the processes generating it occur at national and transnational levels, involving states and multinational corporations with a global reach." Critically, however, O'Lear and Metzo (chapters 4 and 5) argue that such injustice occurs more along the lines of economic status than ethnicity in Azerbaijan and in Tunka. Raudsepp, Heidmets, and Kruusvall (chapter 9) show that environmental injustice in Estonia is perceived in terms of the uneven distribution, not of environmental conditions, but of wealth and of human and property rights.

Seventh, as we mention in our introduction, all FSU republics except the Russian Federation and Uzbekistan ratified the Aarhus Convention. With the assistance of the international NGOs mentioned above in the fifth key point, this convention seems to be helping to shift the discourse to incorporate environmental, public health, and justice issues. As Watters notes (chapter 7):

The Aarhus Convention provides a way for [local] communities to demand their right to be informed, and, more important, their right to have a say in reaching decisions that will affect them. Green Salvation, an Almaty-based NGO, which also actively campaigns at Karachaganak, is the first group to bring a case before the Aarhus Secretariat for noncompliance with the convention (Green Salvation, 2007). This precedent has helped the Berezovka Initiative Group use the Aarhus Convention more effectively in its struggles....

That access to information and public participation are prerequisites for environmental justice is a point echoed by Dominic Stucker (chapter 10). The media play a crucial role in public awareness and involvement, argues Katherine Metzo (chapter 5), who found, however, that industry in Tunka pays newspapers to publish only certain types of articles.

Eighth, Shannon O'Lear (chapter 4) underscores the importance of human security to environmental justice in Azerbaijan:

According to the United Nations Commission on Human Security, human *insecurity*, may become a threat to state stability in instances where, for example, transnational terrorism, environmental pollution, massive population movement and infectious diseases such as HIV/AIDS overwhelm populations and governments alike (Commission on Human Security 2003, 5). Environmental instability and natural resource–related disputes have also emerged as concerns under the label "environmental security" and may have a direct impact on particular populations or regions (see, for example, Ascher and Mirovitskaya 2000).

From our point of view, the human security agenda, like the just sustainability agenda is a middle way between a green, environmental, or conservationist and a social, environmental, or justice agenda. It is a frame that O'Lear further develops: "Those surveyed are predominantly concerned with securing their material well-being and employment, namely, their ability to meet basic needs. Against this background of day-to-day human insecurity, concerns about environmental pollution and the reliability of natural resource–based utilities persist as related strands of environmental justice movement."

Ninth, in a related point, Tamara Steger (chapter 6) argues that "bridging the two levels of environmentalism in Latvia"—the "grass-roots volunteer level...focused on recycling and cleanup campaigns" and the "professionalized and institutionalized level...chiefly addressing pollution problems in a wider policy context"—"would lay the foundation for a discourse on just sustainability by uniting different levels of action."

Finally, Dominic Stucker (chapter 10) provides an interesting take on environmental justice. His work was carried out in Tajikistan according to a Sustainable Livelihoods Framework analysis (Chambers and Conway 1991; DFID 1999; Carney et al. 1999). It focused on the inaccessibility and loss of natural capital, the essential and non-substitutable livelihood asset for rural, agrarian households (Neefjes 2000, 89). Agreeing with Ryan Holifield (2001, 86) that "the definition of environmental justice can vary from place to place and change over time," and proceeding from the central role of "natural capital, the essential and non-substitutable livelihood asset for rural, agrarian households," Stucker notes that his "working definition" of environmental justice "focuses on environmental degradation instead of environmental hazards":

Environmental justice is manifest when every household's right to access and capacity to manage the natural capital necessary to sustain the livelihoods of its members are secured and guaranteed.

Afterthought

There is clearly a wide range of civil society and official state activity throughout the former Soviet Union, from environmental, ecological, or green issues, at one end of the spectrum, to socio-environmental justice issues, at the other. Accordingly, in responding to our first question, we can say that "the extent to which increased popular environmental awareness and associated activism" are "driving public policy and planning" is different in different regions of the former Soviet Union. In some rare cases, they are successful at changing existing planning policy, at least temporarily (e.g., the efforts of RAIPON in the Viliui Sakha regions and of residents in Tunka and Berezovka). In most cases, however, they fail to have the necessary impact to change the policies. "As long as the Russian government continues on a path of centralizing political power, favoring natural resource industries, and controlling information," writes Laura Henry, "Russian greens will have difficulty promoting sustainable development. Without greater access to policy making and stable, transparent, and effective governing institutions, the gap between the rhetoric and the practice of sustainable development is likely to persist."

Similarly, in responding to our second question, the extent to which environmental awareness and activism in the former Soviet Union pursue separate brown and green agendas or a single, middle-way human security or just sustainability agenda varies according to the complex of sociocultural, socioeconomic, political, ethnic, and nationalistic factors that currently define, and are reshaping, the republics. The recent conflict between Georgia and Russia has once again highlighted the complexity and volatility of the relations between the countries of the FSU and has questioned the integrity of the Commonwealth of Independent States. The political climate in and between the states of the FSU will no doubt have a marked effect on the environmental justice and sustainability agendas which are the focus of this book.

References

Ascher, William, and Natalia Mirovitskaya, eds. 2000. *The Caspian Sea: A Quest for Environmental Security*. Boston: Kluwer Academic.

Benford, R. 2005. "The Half-Life of Environmental Justice Frame: Innovation, Diffusion, and Stagnation." In D. N. Pellow and R. J. Brulle, eds., *Power, Justice, and the Environment*, 37–54. Cambridge, MA: MIT Press.

Carney, Diana, M. Drinkwater, T. Rusinow, K. Neefjes, S. Wanmali, and N. Singh. 1999. *Livelihood Approaches Compared*. Department for International Development (DFID). http://www.livelihoods.org/cgi-bin/dbtcgi.exe/ (accessed May 6, 2006).

Chambers, Robert, and Conway, Gordon R. 1991. *Sustainable Rural Livelihoods: Practical Concepts for the 21st Century*. Discussion Paper no. 296. Institute of Development Studies (IDS).

Commission on Human Security. 2003. *Human Security Now*. New York. United Nations.

Department for International Development (DFID). 1999. *Sustainable Livelihoods Guidance Sheets*. London. http://www.livelihoods.org/info/info _guidancesheets.html/ (accessed June 2, 2006).

Green Salvation. 2007. "Herald 2006." Almaty, Kazakhstan.

Holifield, Ryan. 2001. "Defining Environmental Justice and Environmental Racism." *Urban Geography*, 22 (1): 78–90. http://www.bellpub.com/ug/2001/ ad010105.pdf/ (accessed April 5, 2007).

Pellow, D. N., and R. J. Brulle, eds. 2005. *Power, Justice, and the Environment*. Cambridge, MA: MIT Press.

Index

Note: The letter *f* following a page number indicates a figure, *m* a map, *n* a note, and *t* a table.

Urban and Industrial Environments

Series editor: Robert Gottlieb, Henry R. Luce Professor of Urban and Environmental Policy, Occidental College